Understanding Probability and Statistics

Understanding Probability and Statistics *A Book of Problems*

Ruma Falk
with the cooperation of Raphael Falk

The Hebrew University of Jerusalem
Department of Psychology and School of Education
Jerusalem, Israel

A K Peters
Wellesley, Massachusetts

Editorial, Sales, and Customer Service Office

A K Peters, Ltd.
289 Linden Street
Wellesley, MA 02181

Library of Congress Cataloging-in-Publication Data

Falk, Ruma
     Understanding probability and statistics : a book of problems /
Ruma Falk, with the cooperation of Raphael Falk.
          p.   cm.
     Includes bibliographic references and index.
     ISBN 1-56881-018-0
     1. Probabilities—Problems, exercises, etc.   2. Mathematical
statistics—Problems, exercises, etc.   I. Falk. Raphael.
II. Title.
QA273.25.F35   1993
519.2'076—dc20   93-21627
                                                                CIP

Printed in the United States of America
97 96 95 94 93   10 9 8 7 6 5 4 3 2 1

In memory of Leslie V. Glickman (1945–1991).
Friend, teacher, statistician.

Contents

Preface

> We certainly think we know whether we understand something or not;
> and most of us have a fairly deep-rooted belief that it matters.
>
> Skemp (1971, p. 14)

Statistics and probability are often perceived as threats by students, particularly those in the humanities and the social sciences and not so rarely in the natural sciences as well. The very mention of statistics evokes the notorious anxiety associated with mathematics. Furthermore, dealing with uncertainty adds a disconcerting dimension to statistics. When you solve a mathematical problem, hard as it may be, you are rewarded by attaining certainty, but in working out a statistical problem, all your efforts may end up in an average or a probability value which is incapable of telling *what will happen* next. At best, you may learn what will happen *in the long run*, and that too, only in terms of proportions. Some students find this state of affairs quite frustrating.

Without getting into the problem of the roots of math-anxiety and statistics-phobia and the detailed methods of overcoming them, we feel that the best way to allay these fears is by *experiencing* the revelation of *understanding*. One safe way toward this goal is to actively tackle problems concerning the concepts you learn. Nothing is more reassuring than coping with a problem by going back to the definitions of the concepts involved and manipulating them by the rules of the game. One solved problem, worked out on your own, may be worth tens of pages of passive reading.

When some of our more diligent students ask for extra reading, we often discourage these good intentions and recommend instead that they use their energy for independent exploration of problems. Even reading a statistical text is best accomplished when conducted as a problem-solving task. To quote Bush and Mosteller's (1955) word of advice about reading mathematical material:

> Start with a fast reading of the sections concerned to get the general orientation. Then take paper and pencil and work through each derivation step by step. A good understanding of a mathematical development often goes with a feeling that one has invented it oneself, and that the original authors were somewhat opaque (p. viii).

We believe that any serious grappling with a problem, whether you manage to resolve it or merely get to the bottom of the difficulty, is an act of creation. "There is a grain of discovery in the solution of any problem. ... if you solve it by your own means, you may experience the tension and enjoy the triumph of discovery" (Polya, 1957, p. v). It is the creative process of exploration that we hope to offer the student in this collection of problems.

As teachers of statistics, our primary goal in assembling a collection of problems, accompanied by answers and comments, is to positively offer the student the joy of learning and some appreciation of the beauty of probabilistic and statistical thinking. Overcoming the blocks and anxieties is a secondary (though important) goal, hopefully to be accomplished incidentally while achieving the first purpose. It is quite an ambitious expectation. Students who are dismayed by statistics often chuckle at the prospect that some people might enjoy it. Yet, as in every mathematical theory, there is a core of beauty and elegance in the structure of probability theory and in its ability to serve as a model of diverse real-life phenomena. Getting a glimpse of the rewarding aspects of this world may eventually convert a few souls to our camp.

Our selection of problems for this collection was guided by the wish to build up the student's understanding of the *basic concepts* and *principles* of the area. These are not the "plug and chug" variety of problems. Although there may exist good reasons for training oneself to routinely apply standard statistical techniques, this is not what this book is about. Neither is it about data analysis. Our emphasis is not on applications, though we tried to bring examples of varied content areas.

In a way, we are old fashioned by resorting in the computer age, with all the developments in simulations (Gnanadesikan, Scheaffer, & Swift, 1987), computer intensive methods (Diaconis & Efron, 1983; Efron & Tibshirani, 1991), and exploratory data analysis (Tukey, 1977), to the fundamental old concepts. We hold contemporary statistical developments in much respect and admiration, nevertheless, we are convinced that there is no substitute for the prerequisite of internalizing the basic principles.

As a matter of fact, we see some risk in rushing too fast to employ mechanized computerized methods. There is no way to predict when, where, or how lack of understanding of basic principles will reveal itself and what disastrous effects it might entail. Above all, we want to minimize the all too well known phenomenon of a research student running a lot of data through a computer-package, applying all its procedures, and then approaching an expert with his pile of computer outprints, asking her to tell him what the value of kurtosis means ...

The collection encompasses problems of a wide range of levels: starting from high-school level, through that of introductory college courses, up to graduate courses in probability and statistical thinking (to be distinguished from data-analysis and methodological courses). Several problems, which concern fundamental issues, may be instructive for all levels. Others, may serve the teacher by suggesting challenges for class discussions.

Neither calculus nor other techniques of higher mathematics are required for solving the problems. High-school algebra and elementary combinatorics suffice in most cases. Thus, the book is appropriate for *non*mathematicians. However, although most of the problems are technically elementary, not all of them are conceptually trivial, so that the mathematically-inclined student may benefit from them as well.

The problems are roughly arranged, within each chapter, in increasing order of difficulty. What is considered difficult, however, is subjective, and undoubtedly some users of the book would have arranged them differently.

The book is divided into three parts of unequal length. Part I offers problems which require some computation, deliberation, and analysis to answer. There may eventually exist no 'correct' answer to a given problem, and it will remain open for discussion. Part II offers multiple-choice problems. Part III presents some answers.

Part I starts with two sections of descriptive statistics: one on measures characterizing distributions and one on correlation and regression. It is followed by five sections on various topics in probability, which are almost totally devoted to the discrete case. The eighth section offers problems which cut across topics from other sections and a few problems concerning basic concepts in statistical inference. The multiple-choice problems in Part II are divided into three chapters: descriptive statistics, probability, and a last one dealing with the normal distribution, sampling distributions and statistical inference. Numerical tables, such as normal- or t-distribution, are not provided. They should be consulted somewhere else.

Answers are provided to all problems in Part II and to a large part of the problems in Part I. We strongly recommend working out each problem thoroughly on your own before turning to look up the answer. Some of the answers to problems in Part I give just the numerical outcome, others explicate the method of solution as well, and a few include an extensive discussion. We found an outlet for our good advice and instructive comments in these discussions. In a way, they form the 'text' of this book, to be distinguished from the list of problems and solutions (see, e.g., the discussion accompanying the answer to Problem 2.2.12). Problems for which a discussion is offered are marked with an asterisk. Again, we urge you to turn to answers and discussions only after exhausting all efforts to deal with the problem on your own. Give yourself time. Let the problem sink in. You'll probably solve it, but even if you don't, you'll benefit more from the given solution this way.

An extensive list of references includes, not only sources which have inspired composition of our problems, but discussions of related topics as well. These references may help the interested reader to go on exploring some of the issues in more depth and breadth. They may also provide the background for an instructive class discussion or trigger some students' projects.

The problems have been composed and assembled in the course of many years of teaching probability and statistics at the Hebrew University of Jerusalem. Some of the problems are translated from a Hebrew collection.[1] Others were written for examinations and research purposes. Many problems are adapted from different sources in the literature, such as the studies of Tversky and Kahneman on judgment under uncertainty, papers in various educational journals, and a novel by Graham Greene. Many ideas were inspired by colleagues and students. Sources are acknowledged to the best of our ability. We apologize for any oversight that might have occurred in tracing the source of an idea.

Several answers are offered to each problem in Part II (multiple-choice problems). Only one is correct. The others, the distractors, were often chosen on the basis of prevalent beliefs and misunderstandings that we have encountered in class. Some are based on mistakes that appeared in press, as for example, in a paper abundant with false statements on significance testing by Chow (1988). Regrettable as it might be, there is at least one positive aspect to these published mistakes: they provided ideas for several problems, including good distractors.

Our students have contributed to this book over many years, not just by providing us with distractors, but mainly through their comments, inquiries, and surprising insights. The same is true for our teaching assistants. We are grateful to them all.

Our work over the years, which is reflected in this book, was partly supported by the Sturman Center for Human Development, The Hebrew University, Jerusalem.

Ruma and Raphael Falk
Jerusalem, March 1992.

---

[1]Falk, R. (1978). *Problems in probability and statistics.* Jerusalem: Akademon.

Part I

PROBLEMS TO SOLVE

Chapter 1

Descriptive Statistics I

> We look forward to the day when everyone will receive more than the average wage.

> Australian Minister of Labour, 1973
> Quoted from *The College Mathematics Journal*,
> 1987, Vol. **18**, p. 211

1.1 Measures Characterizing Distributions I

1.1.1. The following are measures characterizing the distributions of scores of two parallel English classes:

	class 1	class 2
Mean	78	72
Median	65	73
Standard deviation	16	6

a. In which of the two classes should the teacher invest more time in individual tutoring?

b. In which of the two classes is it more probable to find a few gifted students, that is, outstandingly successful ones?

Justify your answers.

3

1.1.2.   Let a sequence of numbers[1], a_1, a_2, ... , a_n, ... , have the
following characteristic: starting from the third place, each
value is *the arithmetic mean* of all the values that precede it,
that is:
$$a_k = \frac{a_1 + a_2 + \ldots + a_{k-1}}{k - 1},$$

for $k \geq 3$.

What is the nature of this sequence? Explain.

Optional: Prove your assertion formally.

1.1.3.   Students were asked to analyze a set of 50 nonnegative scores,
not all of which were identical. The set included exactly three
0 (zero) scores. It also included two nonzero scores which were
identical to the arithmetic mean of all the scores.

a. One student decided not to include the three zero scores
in his analysis, on the false assumption that zero is not a
number. He correctly calculated the following measures,
based upon his altered data set. For each measure, would
his calculation increase, decrease, or not change the original measure, or is it impossible to tell? *Explain* your
answers.

(1) The arithmetic mean.

(2) The median.

(3) The mode.

(4) The range.

(5) The sum of squares.

b. Another student decided not to include the two scores
that were identical to the arithmetic mean, arguing that
most measures are based on deviations from the mean,
whereas these two scores did not deviate from the mean.

---

[1]Inspired by I. Siwishinsky.

How would the measures that this student obtained change in relation to the correct measures of the complete set of scores? Answer by 'increase,' 'decrease,' 'no change,' or 'impossible to know.' *Explain* your answers.

(1) The arithmetic mean.

(2) The median.

(3) The mode.

(4) The range.

(5) The variance.

1.1.4.   **a.** Write two numbers with mean 10 and variance 4.

b. Write three numbers which form a symmetric distribution with mean 20 and variance $10\frac{2}{3}$.

c. Let a set have two values and variance 9.

(1) What is the range of this set?

(2) What is the mean of the absolute deviations from the median of this set?

d. In a symmetric set there are three values. Their variance is $16\frac{2}{3}$.

(1) What is the range of the numbers?

(2) What is the mean of the absolute deviations from the median?

1.1.5.   Make up a sequence of 8 numbers that will simultaneously satisfy the following requirements:

Mean: 10

Median: 9

Mode: 7

Range: 15.

1.1.6. Twelve observations, $x_1, x_2, \ldots, x_{12}$, are *symmetrically* distributed. Their mean is $\bar{x} = 40$, and their variance is $\sigma_x^2 = 100$.

Another two observations, $x_{13} = 48$ and $x_{14} = 32$, are added to the set. Answer the following questions without calculating the measures characterizing the distribution of the 14 observations.

    a. Would this addition change the *mean* of the distribution? If it does, would the mean increase or decrease?

    b. Would the addition change the *variance* of the distribution? If it does, would the variance increase or decrease?

    c. Would this addition change the *range* of the distribution? If it does, would the range increase or decrease?

    d. Did you use the symmetry of the distribution in answering questions **a** to **c**? If you did, in which of them?

1.1.7.   **Minimize Your Losses**[2]

You are given the following set of seven numbers:

$$4, \ 3, \ 2, \ 10, \ 6, \ 14, \ 10.$$

Let us play a game with five parts. In each part you will choose a single number to represent the set of seven numbers.

You will get an initial sum of 170 points, and in each part you will use a different rule to calculate the disparity between the set's values and your suggested value. This disparity is your fine, and you will deduct it from your total score.

For each case, write the *representative value* that you suggest (give a number) and calculate your *fine*.

The number of points left over at the end will be your score for the game. Try to maximize it.

---

[2]See Falk (1980).

    a. (1) Pay the sum of the distances (absolute differences) between your representative value and the values of the set.

    (2) Pay one point for every element in the set which does not equal your number.

    (3) The fine equals the maximal absolute deviation.

    (4) Pay the absolute value of the sum of all deviations from your representative value.

    (5) The fine equals the sum of the squared deviations between your representative value and the values of the set.

    b. How many points do you have left?

1.1.8.   Minimize Your Losses — General[3]

A student is presented with the following game. She is given a sequence of numbers, x_1, x_2, ..., x_n, not all of which are identical ($n > 2$). She is asked to examine the numbers and then to suggest one value to represent them all.

To start with, the student is given a sum of money. At every stage the disparity between her suggestion and the given numbers is checked. She has to pay fines which depend on the deviations between the sequence and her suggestion, and are calculated according to some given rule.

Write which value she should suggest in each case, so that at the end of the game she has the largest possible sum of money.

    a. For each i, she has to pay as many dollars as the distance (absolute deviation) between her suggested value and the actual value of x_i.

    b. She pays the sum of the squared deviations between her number and each x_i.

---

[3]See Falk (1980).

    c. She pays a fine equal to the absolute value of the sum of all deviations.

    d. She pays \$1 for each x_i which does not equal her number.

    e. A '+' is scored for each positive deviation and a '−' for each negative one. She pays a fine equal to the absolute difference between the number of +'s and the number of −'s.

    f. Her payment equals the largest absolute deviation between her number and an x_i.

    g. Her payment equals the product of the absolute values of the deviations. (Please, watch out!)

1.1.9. Five houses are located along Maple St., a straight east-west road. Their distances in kilometers from the west end of the street are as follows:

$$7,\ 2,\ 1,\ 7\frac{1}{2},\ \frac{1}{2}.$$

Where, along the street, would you place each of the following establishments? Indicate each location by giving its distance from the west end of the street.

    a. A mailbox that would serve the 5 houses in the most efficient way. That is, the sum of the walking distances from the houses to the mailbox should be minimal.

    b. A north-south main road, intersecting Maple St. at a right angle and dividing it into 'west' and 'east' sections. The sum of the driving distances from the houses in the east section to the road should equal the sum of the driving distances from the houses in the west section to the road.

    c. An emergency siren. The siren should be heard in all the houses, including the most distant ones. Obviously a minimal volume siren is desirable.

1.1.10. Make up three pairs of numbers, (x_i, y_i) for $i = 1$, 2, 3, representing the incomes of three fictional married couples. Let x_i stand for the income of husband i and y_i for that of his wife. Choose your numbers so that the median of the total incomes of the couples is *different* from the sum of the medians of the incomes of the husbands and the incomes of the wives.

In symbols, your example should satisfy:

$$Me(x + y) \neq Me(x) + Me(y).$$

1.1.11. In an educational research project it is necessary to construct a control group that shares some features with the experimental group. There are 12 children in the experimental group, and the investigator decides that the control group should be the same size.

The mean and the variance of the variable x in the experimental group are 6.0 and 14.00, respectively. The investigator wishes to construct the control group so that both groups will have the same mean and variance.

After ten children have been selected for the control group, the mean of their x-values is 5.8 and the variance is 16.36. Two additional children will be selected for the control group.

What should be the x values of these children, so that the experimental group and the control group will have equal means and equal variances?

1.1.12. You are given three tables, each presenting the *cumulative* frequency distribution of a discrete variable. There are 100 observations in each distribution.

In each case, find the *mean* and the *variance* of the distribution.

x	$F(x)$
0	0
1	0
2	0
3	100

a.

x	$F(x)$
0	50
1	50
2	100

b.

x	$F(x)$
1	25
2	50
3	75
4	100

c.

1.1.13. Eight people took a test in which one can score only 1, 2, or 3.

    a. You know that exactly two people scored 1 and that the distribution is *symmetric*. What is the variance of the set of scores?

    b. Let the variance of the set be 1. List the eight scores.

    c. Given that the mean of the scores is 3, what is the standard deviation of the set of scores? [4]

1.1.14.   **a.** The mean score of 15 students was 82.00. Another student, whose score was 93, joined the group.

      Compute the mean score of all 16 students (accurate to two decimal places).

---

[4]Inspired by a somewhat different problem in Freedman, Pisani, and Purves (1978, p. 52, problem 2).

b. Let $\underline{M_n}$ be the mean of $\underline{n}$ scores $(x_1, x_2, \ldots, x_n)$. Let $\underline{x_{n+1}}$ denote a new score that is added to the group. Give a formula for computing the mean, M_{n+1}, of all $n+1$ scores, *using the above underlined symbols*.

c. Let d denote the *difference* between the mean M_{n+1} (of the $n+1$ scores) and the mean M_n (of the n scores):

$$d = M_{n+1} - M_n.$$

Use *the symbols underlined in paragraph* **b** to express d in a simple way.

d. When does the addition of a new score x_{n+1} to the existing group of n scores result in an increase of the group's mean, when does it result in a decrease of the mean, and when does the mean stay unchanged? Explain briefly, referring to your formula for d.

1.1.15. The treasury department is considering several schemes for revising its salary and employment policies for government workers.[5]

The following three schemes are suggested. Determine, in each case, how the suggested revision would affect each of the following measures:

(1) The *mean* monthly salary in dollars.

(2) The *variance* of the monthly salaries.

(3) The *standard deviation* of the monthly salaries.

(4) The *median* monthly salary.

(5) The *modal* monthly salary.

a. Each employee will get a raise of $125 per month.

---

[5]Adapted from Freedman et al. (1978, p. 64, problem 7).

  b. The salaries will be increased by 15% across the board.

  c. The number of employees at each salary level will be decreased to 90% of their original number.

1.1.16. Some years ago a well-known public official left California and moved to Alabama. A local California reporter revealed both his regional chauvinism and his feelings about the official when he remarked that "on this occasion he raised the mean IQ in both states."[6]

  a. Still, such a statistical change is theoretically possible. Explain how.

  b. Considering this, and since the mean IQ of the US is the average of the means of all states, it appears that by a mere reshuffling of populations between states, one could increase the mean IQ of the US! Is this so? Explain.

1.1.17. The curve on the next page represents the *cumulative* frequency distribution (in percentages) of a continuous variable x whose minimal value is A and maximal value is B.

  a. Use these data to draw (schematically) a density curve, $f(x)$, on the x axis from A to B.

  b. Insert the missing sign, $<$, $=$, or $>$ (one and only one), between the symbols on the next page, all of which refer to the distribution described in the curve.

---

[6]See Falk and Bar-Hillel (1980).

(1) Me $\bar{x}$.

(2) $(A + B)/2$ Me.

(3) $F(\bar{x})$ 50%.

(4) $F(Q_3) - F(Q_1)$ 50%.

(5) Q_1 $A + \dfrac{B - A}{4}$.

(6) $F(Q_3)$ 75%.

(7) $Q_1 - A$ $B - Q_3$.

(8) $Q_3 - Me$ $Me - Q_1$.

(9) $F(Q_3) - F(Me)$ $F(Me) - F(Q_1)$.

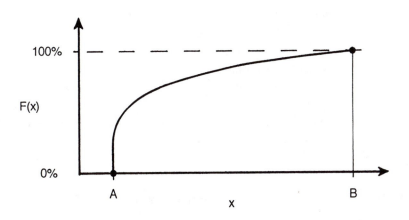

1.1.18. Demographers wish to study a small village, comprised of only 30 families.[7]

They plan to visit all 30 families and record the number of children in each. They will then use these numbers to calculate the mean number of children per family. Let's call this plan *Method A*.

When it turns out that all the children are of school age, and all attend the same single village school, *Method B* is sug-

---

[7]Based on an article by Madsen (1981).

gested: the researchers will go to the school and ask *each child* for the number of children in his/her family (the questioned child included) and then calculate the mean of the numbers obtained.

The following is the distribution of families by the number of their children:

Number of children	i	0	1	2	3	4	5	6	7	Total
Number of families	f_i	3	4	6	6	4	3	2	2	30

a. Calculate the mean by *Method A*.

b. Calculate the mean by *Method B*.

c. What is the meaning of these two means? Namely, what *question* does the calculated result answer in each of the two methods?

d. Try to invent an example of a distribution of n families ($n > 2$) by number of children, so that the relation between the size of the means obtained by *Method A* and by *Method B* would be reversed (relative to your answers to a and b). If you succeed, show that the order is indeed reversed; if not, explain why.

1.1.19.   **Which Mean Do You Mean?**[8]

In parts a and b(1) of this problem, you are asked to suggest appropriate averages (or means) for the situations described.

a. Suppose that a company has sales of $100,000 in its first year of operation, and that for the next five years its sales increase by 10 percent, 70 percent, 25 percent, 10 percent, and 15 percent, respectively (see Lubecke, 1991). How would you best average the yearly increase in sales over these five years?

---

[8]This title is borrowed from two, apparently independent, papers: by Burrows and Talbot (1986), and by Lubecke (1991).

b. (1) A skier rides a lift to the top of a mountain. It moves upwards at the speed of 5 kilometers per hour. On the way down, he skies at the speed of 15 kilometers per hour. What is his average speed for the total trip?

(2) Another skier, too impatient with the slow ride up at 5 kilometers per hour, expresses his wish to travel downwards so fast that he will raise his average speed to 10 kilometers per hour for the round trip up and down the slope. How fast must he ski down?[9]

1.1.20. Given a set of numbers, x_1, x_2, ..., x_n, prove that the sum of the absolute deviations is minimal when the deviations are taken from the *median* of the set (see Problems 1.1.7.**a** (1), 1.1.8.**a**, and 1.1.9.**a**).

1.1.21. How Far Apart Can the Mean and the Median Be?

It is well known that the arithmetic mean of a set of values and its median need not coincide. However, the limit on the possible gaps between these two averages is not as well known.

Let $x_1 \leq x_2 \leq x_3 \leq \ldots \leq x_n$ be a set of numbers with nonzero standard deviation σ. Prove that $|M - Me| \leq \sigma$, where M denotes the arithmetic mean and Me the median of the set.[10]

1.1.22. Two basic measures of dispersion are the standard deviation σ and the range R. Prove that for any finite distribution, $\sigma \leq R/2$.

---

[9] Adapted from a puzzle by Gardner (1982, p. 142).

[10] See Book and Sher (1979) and Falk (1981) for proofs of this theorem. A more extensive and refined discussion of this problem (and related ones) can be found in Page and Murty (1982, 1983).

1.2   Correlation and Linear Regression

1.2.1.   In each part of this problem find the missing numbers (x and/or y values) which satisfy the requirements listed next to each table. No elaborate computations of correlations (or other measures) are needed to complete any of the solutions.

a.

x	y
1	7
2	12
6	32
10	
	67

$r_{xy} = 1$

b.

x	y
1	
2	
3	
4	
5	

$r_{xy} = -1$

c.

x	y
2	
3	
7	
12	
20	

$\sigma_y^2 = 9\sigma_x^2$

$r_{xy} = 1$

d.

x	y

$\bar{x} = \bar{y}$

$\sigma_x = \sigma_y \neq 0$

$r_{xy} < 0$

1.2.2.   The dexterity of five people's left and right hands was measured, and the following scores were obtained:

Subject	Left hand score	Right hand score
A	4	6
B	1	3
C	5	7
D	3	5
E	10	12

Answer the following questions, without performing calculations.

    a. What is the linear correlation coefficient (r) between the scores for the right hand and those for the left hand?

    b. What is the (Spearman) rank-order correlation coefficient (r_s) between the scores for the right hand and those for the left hand?

    c. After a while, another subject (F) was tested and given the following scores: left hand 9, right hand 8. Will this additional datum affect the linear correlation coefficient r or the rank-order correlation coefficient r_s, and how?

1.2.3.   Consider the following n pairs of observations:

x	y
x_1	y_1
x_2	y_2
.	.
.	.
.	.
x_n	y_n

a. A statistician observed that all the y values were even. Consequently she divided each by 2.

(1) Would the linear correlation coefficient between the two variables change as a result of this operation, and how?

(2) Would the rank-order correlation coefficient between the two variables change as a result of this operation, and how?

b. Another statistician performed the following transformation on the x values: each x was squared; namely, it was changed to the variable $x' = x^2$ (all x values were positive).

(1) Would you expect the linear correlation coefficient between the variables x and y to be equal to, or different from the linear correlation coefficient between x' and y?

(2) Same question with respect to the rank-order correlation coefficient between x and y.

1.2.4. The following are two values of the correlation coefficient between x and y:

(1) r_1 $=$ 0.7.
(2) r_2 $=$ -0.9.

a. In which of the two cases is our ability to linearly predict y from x higher?

b. What percentage of the variance in y values can be accounted for (by knowledge of x) in each case?

c. What percentage of the variance in x values can be accounted for (by knowledge of y) in each case?

1.2.5. In the department of psychology at a certain university the studies are divided into two parts: theoretical and practical. Students' scores in the theoretical part are denoted x, and those in the practical part are denoted y.

The coefficient of correlation between the two scores is $r_{x,y} = 0.65$, and the mean of the theoretical scores is $\bar{x} = 75$.

One student took only the theoretical studies and obtained a score of 83. He postponed his practical studies to another year. The university's statistician used regression to predict the score the student would have obtained in the practical part. She sent that predicted score to the research department.

Due to a misunderstanding, the research people took that score to be the student's *actual* score in the practical part; they thought they were supposed to use that score in order to predict the student's score in the theoretical part. Hence, they used the regression principle to predict the student's x score, based on the y score obtained from the statistician.

What will the value of that predicted theoretical score be?

(1) Greater than 83.

(2) Equal to 83.

(3) Impossible to know because of insufficient data.

(4) Less than 83 and greater than 75.

(5) Equal to 75.

(6) Less than 75.

Explain your choice.

1.2.6. Given on the next page are seven pairs of formulas that are supposed to be *linear regression equations*. The variable $\hat{y}$ denotes the regression of y on x, and $\hat{x}$ denotes the regression of x on y.

In each case, decide whether the given pair of functions could or could not represent a pair of regression lines.

a. $\hat{y} = 3x + 5$; $\hat{x} = 2y - 3$.

b. $\hat{y} = 3$; $\hat{x} = 5$.

c. $\hat{y} = 1.5x + 1$; $\hat{x} = -0.5y + 0.8$.

d. $\hat{y} = 3x^2 + 4$; $\hat{x} = 0.2y^2 + 1$.

e. $\hat{y} = 2x + 4$; $\hat{x} = 0.5y - 2$.

f. $\hat{y} = 4$; $\hat{x} = 1.2y - 3$.

g. $\hat{y} = 2x + 2$; $\hat{x} = 0.5y + 8$.

1.2.7. Psychologists composed an intelligence test and ran it on a large group of children with a wide age range.

The linear correlation coefficient between test score and age was 0.61.

The mean of the ages was 11.6, and the standard deviation was 2.72.

The mean score was 79.5, and the standard deviation of the scores was 9.23.

It was decided to use the test results in order to determine a child's '*mental age*.' Starting with a child's test score, the linear relations obtained between test scores and the children's ages was used to fit a 'mental age' to the child. A gifted 9 years old child scored 81 on the test. What mental age will this method assign the child?

Chapter 2

Probability I

> While the individual man is an insoluble puzzle, in the aggregate he becomes a mathematical certainty. You can, for example, never foretell what any one man will do, but you can say with precision what an average number will be up to. Individuals vary, but percentages remain constant. So says the statistician.
>
> Sherlock Holmes in *The Sign of Four*, 1890.

2.1 Events, Operations, and Event Probabilities

2.1.1. Determine whether each of the following assertions is true for every A, B, and C in a sample space Ω.

  a. $\overline{A \cup B} = \overline{A} \cup \overline{B}$.

  b. $A \cap \overline{B} \cap C \subseteq A \cup B$.

  c. If $A \cup B = A \cup C$ then $B = C$.

  d. If $A \cap B = A \cap C$ then $B = C$.

  e. $A \cup (B \cap C) = (A \cup B) \cap (A \cup C)$.

  f. $\overline{A - B} = \overline{A} \cup B$.

  g. $A - B \subseteq \overline{B}$.

  h. $A - \overline{A} = \phi$.

  i. $\overline{A \cup B \cup C} \subseteq \overline{A \cap B \cap C}$.

j. If $A \cup B = A \cap B$ then $A = B$.

k. $\overline{A \cap \overline{A}} = \Omega$.

l. $A = (A \cap B) \cup (A \cap \overline{B})$.

2.1.2. Let Ω be the set of all students at a given university.

Several subsets of Ω are represented below. Some of the sets are represented by symbols and some by the probability that a random student will belong to the set.

Use the given symbols and probabilities to complete the missing items in the table.

Definition of set in words	Symbol	Probability
All social-science students	S	
All research students	R	
All social-science and/or research students		0.25
All research students who are not from social sciences		0.05
All social-science students who are not research students		0.10
All social-science-research students		

2.1.3. Here is a rough description of an experiment carried out by psychologists:[1]

Subjects were given the following judgment problem: "Think of a population of women with academic degrees in the social sciences. Consider a random woman from such a population. Rank order the following categories according to the probability that the woman will belong to them."

---

[1]Inspired by Tversky and Kahneman (1983).

    a. Employed at a university.

    b. Married and unemployed.

    c. Owns her own business.

    d. Unemployed.

The subjects' rankings, from the most probable possibility to the least probable one, matched the order in which the categories are written, that is, **a** was rated most probable and **d** least probable.

Are these rankings compatible with the laws of probability theory, or do they violate them in any way? Explain. (Note that you are not being asked about the state of employment of academic women. Rather you are asked to evaluate the consistency of the judgments of the experimental subjects.)

2.1.4.   **A Domestic Dispute**

Tom and Harriet cannot agree on whether to go to the baseball game (Tom's preference) or to the movies (Harriet's choice). Flipping a coin and deciding for Harriet if 'heads' and for Tom if 'tails' appears too trivial. They discuss several chance procedures for making their decision.

Let T be the event 'Tom wins' and H the event 'Harriet wins.' Find $P(T)$ and $P(H)$ for each of the suggestions described below, and determine whether each procedure is fair or biased.

    a. Playing one round of "Scissors, Paper, Rock"[2] to determine whose wish will be granted.

    b. Flipping 2 coins. Harriet prevails if at least one outcome is heads, Tom — otherwise.

---

[2]The two players simultaneously signify with their hand, through conventional symbols, one of three objects: Scissors (two outstretched fingers), Paper (an open palm of hand), or Rock (a closed fist). If both signify the same object, the game is a draw. Otherwise, superiority is based on the fact that Scissors cut Paper, Rock breaks Scissors, and Paper covers Rock (Williams, 1966, p.98).

c. Giving each a box containing three notes numbered 1, 2, and 3, and having each blindly draw one of the notes. If the sum of their draws is even — Tom wins, if it is odd — Harriet wins.

d. Rolling two dice and computing the *absolute difference* between the two numbers obtained. Tom wins if the outcome is either 1 or 2, Harriet — otherwise.

2.1.5. A and B are *disjoint* events in the probability space (Ω, P). Let $P(A) = a > 0$; $P(B) = b > 0$.

Calculate the following probabilities, using the values a and b:

a.	$P(A \cup B)$.	**e.**	$P(\overline{A \cap B})$.
b.	$P(A \cap B)$.	**f.**	$P(\overline{A} \cap \overline{B})$.
c.	$P(\overline{A} \cup B)$.	**g.**	$P(A - B)$.
d.	$P(\overline{A} \cap B)$.		

2.1.6.   Let A and B be events in a probability space (Ω, P) that satisfy $A \subseteq B$. Let $P(A) = a > 0$, and $P(B) = b$. We know that $a \leq b$. (Why?)

Calculate the following probabilities, using the values a and b (Venn diagrams may be helpful).

a.	$P(A \cup B)$.	**d.**	$P(\overline{A} \cap B)$.	**g.**	$P(\overline{A} \cup B)$.
b.	$P(\overline{A})$.	**e.**	$P(A - B)$.	**h.**	$P(\overline{A} - B)$.
c.	$P(A \cap B)$.	**f.**	$P(\overline{A \cup B})$.	**i.**	$P(\overline{A} - \overline{B})$.

2.1.7. A and B are two events in a probability space. You know that $P(A) = 0.20$ and $P(B) = 0.40$. Find the probability that *at least one* of the events A or B will occur, under each of the following conditions:

    a. $P(A \cap B) = 0.15$.

    b. A and B are disjoint events.

    c. A is (properly) included in B (i.e. $A \subset B$).

    d. $P(B - A) = 0.35$.

    e. $A \subset \overline{B}$.

    f. $\overline{B} \subset \overline{A}$.

2.1.8.   **How to Secure Hot Water in Your Home?**[3]

Abraham and Sarah are a married couple. They have an electric boiler for heating water in their home. The boiler is operated by pressing a switch once; it is turned off by pressing the same switch again.

Consider a randomly-chosen winter day. Let A denote the event 'Abraham pressed the switch,' and S the event 'Sarah pressed the switch.' Let H denote the event 'There is hot water in the boiler.'

    a. Use the appropriate operations to express the event H in terms of events A and S.

    b. The probability that Abraham will remember to press the switch is 0.7 and that of Sarah remembering to do so is 0.4.

    Compute the probability of H:

        (1) Assuming independence between each of the spouses remembering (or forgetting) to press the switch and

---

[3]Based on an idea suggested by Tali Arbel-Avishai.

lack of communication between them. (For the definition of independence, see Problem 2.3.15 and the answer to that problem.)

(2) Assuming that Abraham and Sarah remember (or forget) to press the switch independently of each other, but there is a red bulb that lights up above the switch whenever the heating is on.

2.1.9. Complete the missing probabilities[4] in the table below.

$P(A)$	$P(B)$	$P(A \cup B)$	$P(A \cap B)$	$P(A - B)$	$P(B - A)$	$P(A \triangledown B)$
0.7	0.6	0.9				
			0.1	0.3	0.2	
	0.8	1	0			
0.6	0.4		0.4			
0.5	0.5	0.5				
	0.6		0.2	0.1		
		0.5	0	0.3		
		0.8		0.4	0.1	
0	0.3					
1	0.4					
0.8			0.5			0.4
0.4		0.7				0.5

2.1.10.   **A Plausible Paradox in Chances**[5]

This problem's title is taken from that of a note by Francis Galton which was published in *Nature* in 1894. Here is the 'paradox' in Galton's words:

---

[4] $A \triangledown B$, at the head of the last column, denotes the symmetric difference of A and B, that is, $(A \cup B) - (A \cap B)$ (see the answer to Problem 2.1.8a).

[5] We owe many thanks to the late Leslie Glickman for referring us to this 'paradox.'

The question concerns the chance of three coins turning up alike, that is, all heads or else all tails. The straightforward solution is simple enough; namely, that there are 2 different and equally probable ways in which a single coin may turn up; there are 4 in which two coins may turn up, and 8 ways in which three coins may do so. Of these 8 ways, one is all-heads and another all-tails, therefore the chance of being all-alike is 2 to 8 or 1 to 4.

Against this conclusion I lately heard it urged, in perfect good faith, that as at least two of the coins must turn up alike, and as it is an even chance whether a third coin is heads or tails; therefore the chance of being all-alike is as 1 to 2, and not as 1 to 4. Where does the fallacy lie? (Galton, 1894, p. 365)

Though it is clear that the first reasoning is right and the second is wrong, pinpointing the exact source of error in the second is not that simple. Can you do it?

2.2   Problems Involving Combinatorics

Games of chance are probably as old as the human desire to get something for nothing.

Lightner (1991, p. 623)

2.2.1.   A soccer lotto form consists of a table of 13 rows, corresponding to 13 games, and three columns headed 1, 2, and X. One predicts the outcomes of the soccer matches by marking one column in each row, with 1, 2, and X representing victory for team 1, victory for team 2, and a draw, respectively. A person who does not know a thing about soccer fills a lotto form by *guessing*, namely by putting down 1, 2 or X with equal probabilities for each of 13 games.

a. What is the probability that such a person correctly guesses the outcomes of all 13 games?

b. What is the probability that the person correctly guesses the outcomes of at least 12 games?

2.2.2. A school teacher is aware that she has problems remembering names. However, she cares about her pupils and she tries very hard to memorize all their names.

One day four new girls are admitted to her class. Their names are Alice, Barbara, Cynthia, and Doris. The teacher makes a note of their four names, and she rehearses them all evening at home. The next day, when she sees the new students again, she addresses them each by name, although she feels that she has no idea which name belongs to which girl.

What is the probability that *no one* of the girls will be called by her own name? (See Hadar & Hadass, 1981b, for an extension of this problem to n items and an analysis of the typical difficulties involved in reaching a solution).

2.2.3.   **The Tea Tasting Lady**

Here is an experiment concerning the claim of a renowned British lady, described in Fisher's[6] words:

> A lady declares that by tasting a cup of tea made with milk she can discriminate whether the milk or the tea infusion was first added to the cup. We will consider the problem of designing an experiment by means of which this assertion can be tested. ... Our experiment consists in mixing eight cups of tea, four in one way and four in the other, and presenting them to the subject for judgment in a random order. The subject has been told in advance of what the

---

[6]Fisher (1960, p. 11).

test will consist, namely that she will be asked to taste eight cups, that these shall be four of each kind, and that they shall be presented to her in a random order. ... Her task is to divide the 8 cups into two sets of 4, agreeing, if possible, with the treatments received.

Suppose the lady has no capacity for discrimination and that she only *guesses*. What is the probability that she will correctly identify all the cups?

2.2.4. Eight persons attend in a party.

    a. At the end, three guests are chosen by lottery to receive first prize, second prize, and third prize gifts. What is the probability that Abe will win the first prize, Bill the second prize, and Charles the third prize?

    b. At the end, three of the eight guests are chosen by a random draw and are handed three identical prizes. What is the probability that Abe, Bill, and Charles will be the winners?

2.2.5. * A young Israeli girl gave a soldier her phone number during a hurried parting. The soldier remembers the 6 digits of her

---

* A reminder: the asterisk signifies that the answer to the problem includes some discussion or extension.

telephone number, but in his emotional state he *does not re-member the order* of the digits. He has in his possession only one token and is going to try to dial the six digits in some *random* order. (In Israel, phone numbers have six digits, and tokens are used to pay for calls from public phones.)

What is the probability that he will get to the right girl on this only trial, if the figures are:

  a. 4, 2, 3, 6, 1, 5.

  b. 4, 4, 4, 6, 1, 5.

  c. 4, 4, 4, 6, 6, 6.

2.2.6. A historical anecdote from the seventeenth century (see Freedman et al., 1978, pp. 221–222 and Kreith & Kysh, 1988) tells about Italian gamblers who used to bet on the total number of dots rolled with three dice. They believed that the chance of rolling a total of 9 ought to equal the chance of rolling a total of 10. They figured that there are altogether six ways to obtain the sum 9 in three dice:

1 2 6,   1 3 5,   1 4 4,   2 3 4,   2 2 5,   3 3 3.

Similarly, they found six ways to obtain 10:

1 4 5,   1 3 6,   2 2 6,   2 3 5,   2 4 4,   3 3 4.

Thus, they argued that 9 and 10 should have the same chance.

However, experience showed that 10 came up a bit more often than 9. The gamblers asked Galileo (1564–1642) for help with the apparent contradiction, and he resolved the 'paradox' for them. Can you do the same?

2.2.7. A teacher gave his students 20 problems in order to study for a test. Five of the 20 problems were chosen at random for the actual test. A student managed to study only a certain set of five problems.

    a. What is the probability that exactly these five problems were on the test?

    b. What is the probability that none of the five problems that the student studied were on the test?

2.2.8.   **Birthday Coincidences**[7]

Ten people get together on New Year's Day. The question of coincidences soon comes up. They wonder about the following:

    a. What is the probability that at least two of them share the same birthday?

    b. What is the probability that at least one person celebrates his or her birthday on that very day (January 1)?

Compute the two probabilities, assuming no leap year, independence between these people's birthdates, and that each date is equally likely for a birthday.

2.2.9.   Five different kinds of plastic toys are hidden in boxes of cornflakes: a dog, a duckling, a cow, a horse, and a farmer.

The toys are distributed at random, one per box, from a huge stock containing equal amounts of the five prizes. A little boy

---

[7]For extensions and elaborations on the birthday problem, see Diaconis and Mosteller (1989), Hocking and Schwertman (1986), and Mosteller (1965, pp. 46–51).

wishes to collect a whole set of toys. His parents buy a new box of cornflakes each week.

    a. What is the probability that the boy completes the set within the first five weeks ?

    b. If the boy finds a cow and a horse in the first two boxes, what is the probability that the next three boxes complete the set?

2.2.10. Nursery-school children are given English letters printed on cards, and are allowed to play with them to become familiar with the symbols.

    a. One toddler plays with 10 cards bearing the following letters:

$$\text{A B C K L M N U V W.}$$

After much turning and shuffling, he randomly picks four letters and arranges them in line.

What is the probability that the cards spell out the word LUCK? (Would you consider such an event 'lucky'?)

    b. Another little girl plays with the following 11 cards:

$$\text{C C C D E E I I N N O.}$$

After many manipulations, she randomly orders all of them in line. How probable is the coincidence that she spells the word COINCIDENCE?

2.2.11.   **Enjoy Your Ice Cream**[8]

Thirty-one flavors are offered in an American ice cream parlor. A serving consists of three scoops of ice cream. A tourist who

---

[8]We thank Kurt Kreith for the idea for this problem (see Kreith, 1992).

doesn't speak English cannot communicate with the salespeople. Since the line is long and everybody is in a hurry, he resigns himself to getting whatever he's given.

a. The three scoops are heaped in a cone, one on top of the other. The tourist would have liked vanilla, strawberry, and coffee flavors, in that order, from top to bottom.

What is the probability that his exact wish is granted if

(1) The salesclerk instructs the computer to randomly select one flavor of the 31, and to do it independently for the bottom, middle, and top scoop?

(2) The salesclerk randomly picks the bottom, middle, and top scoop (she would thus not pick a flavor more than once)?

b. The three scoops are placed beside each other in a cup so that one can eat them in any order one wishes. The tourist would have liked vanilla, strawberry, and coffee flavors.

What is the probability that his exact wish is granted if

(1) The salesclerk instructs the computer to select three flavors so that all the different (unordered) triplets — allowing repetitions of a flavor — are equally likely?

(2) The salesclerk randomly picks three flavors out of the 31 (she would thus not pick a flavor more than once)?

Compare this problem with Problem 2.2.12.

2.2.12. * In an engaging article on "Randomness in Physics and Mathematics," Troccolo (1977) presents an analysis of an "absurd" problem yielding insights into atomic physics. In that problem, the host, a physicist named Dr. Zed, poses a probability problem to his three guests, Professors Alpha, Beta, and Gamma, all mathematicians. Each mathematician interprets the problem somewhat differently, by his own assumptions concerning the random mechanism behind the story, and thus three different answers are given.

Here is Dr. Zed's problem:

> Gentlemen, I have noticed you admiring the five miniature hand-tooled leather boxes that are sitting on my coffee table. I acquired them on a visit to India and am rather fond of them. There is a little problem associated with them that I think might amuse you. Before you came, I asked my wife to take three coins from her purse and distribute them at random in the boxes. This means, of course, that any possible arrangement has an equal chance of being the one selected. The problem is, gentlemen, what is the probability that the first box is empty? (Troccolo, 1977, p. 772)

From here on, we offer a variation of Troccolo's story. We take the liberty of changing roles between the guests and inviting to the party a fourth mathematician, Professor Delta. We present each professor's assumptions. In each case determine the probability that the first box is empty contingent on the professor's assumptions.

a. *Professor Alpha:* It stands to reason that Mrs. Zed has chosen three different small coins, like a penny (P), a dime (D), and a nickel (N). That way she would be able to fit either one or more than one coin into any box of her choice. Furthermore, we would be able to distinguish between an arrangement like

P		D	N	

,

and one like

D		N	P	

.

    b. *Professor Beta*: Indeed, I agree that Mrs. Zed would have chosen three distinguishable small coins, however, I do *not* believe she would have placed more than one coin in any one box. There are enough boxes available for her to use a given box no more than once.

    c. *Professor Gamma*: Surely Mrs. Zed would have used three identical coins, like three dimes. Hard to believe she would use three different coins. She would, however, presumably choose small coins (like dimes) so that she could fit one, two, or even three coins into the same box.

    d. *Professor Delta*: My guess is that Mrs. Zed has selected three half-dollars (i.e., identical large coins) to put in the boxes. However, as attractive as these boxes are, they are so small that there can be no more than one coin in any one box.

2.2.13. Five married couples are invited to a party. At the entrance, each guest randomly draws one of 10 notes to determine his or her place in a *row* of ten seats.

Calculate the probability of the following events:

    a. Along the row, the ten people will sit alternating by sex.

    b. The five women will sit next to each other at one end of the row, as will the five men.

    c. Each man will sit next to his wife.

2.2.14. What is the probability that a randomly selected 5-digit number reads the same backward and forward[9]? (50205 is one example of such a number, called a palindrome.)

Assume that each of the 10^5 ordered quintuplets of digits has equal probability of being selected.

[9]Robyn Dawes (personal communication, 1980).

2.3   Conditional Probabilities and Dependence/Independence between Events

2.3.1.   Consider the following limerick composed by Sir Arthur Eddington:[10]

> There once was a brainy baboon
> Who always breathed down a bassoon,
>   For he said, "It appears
>   That in billions of years
> I shall certainly hit on a tune."

The verse contains 118 letters (ignoring blank spaces and punctuation marks). Suppose we randomly select one letter from the sequence of 118 letters.

Let T denote the event that this letter is a *t*, and let *1st* denote the event that this letter begins a word.

Compute the following probabilities by counting: $P(T)$ and $P(T \mid 1st)$.

Which characteristic of the English language is expressed by the relationship between these probabilities?

2.3.2.   Consider the following expression of Charlie Chan, quoted by Gardner:[11]

> Strange events permit themselves the luxury of occurring.

This sentence comprises 49 letters (disregarding blank spaces). Suppose we randomly select one letter from the sequence of 49 letters. Let H be the event that this letter is an *h*. Let

---

[10]Quoted in Kasner and Newman (1949, p. 223).
[11]Gardner (1957, p. 307).

$P(H \mid T)$ denote the conditional probability that the letter is an h given that it is preceded by a t.

Compute the following probabilities by counting: $P(H)$ and $P(H \mid T)$.

Which characteristic of the English language is expressed by the relationship between $P(H \mid T)$ and $P(H)$?

2.3.3. Dutch children often use the procedure described below in order to determine which of six children will play first.

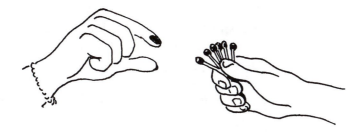

Six identical-looking matches are used. An uninvolved person secretly breaks one match at its lower end and then holds all six in his or her palm, so that the lower ends are hidden, and only the six upper ends are visible.

The first child randomly draws a match. If the match is the broken one, the child is selected, and the drawing procedure stops. If not, the second participant draws a match, and so on. The drawing procedure terminates when the broken match is selected.

Let us denote the six children by their ordinal numbers in the draw: 1, 2, ..., 6. For each child, compute the probability that he or she will be selected to open the game.

Is the drawing procedure *fair* or not? Explain (also see Problem 2.4.12).

2.3.4.   The following demographic data and questions concern *men* in a certain country.

    a. For every 1000 live births, the number of those who survive to their 65th birthday is 746 (in the common notation: $l_{65} = 0.746$).

    The probability that a man who just turned 65 will die within five years is 0.160.

    What is the probability that a baby boy born in this population will reach his 70th birthday? That is, $l_{70} =$?

    b. For every 1000 live births, the number of those who survive to their 50th birthday is 913 ($l_{50} = 0.913$), and the number who survive to their 55th birthday (out of the same 1000 births) is 881 ($l_{55} = 0.881$).

    Calculate the probability that a man just turning 50 will die within five years (i.e., the *conditional probability* that a man who has reached his 50th birthday, will not reach his 55th birthday).

    c. When the data in paragraph **b** were presented to a class of statistics students, one of them was alarmed to learn that the probability of men surviving to age 50 was 0.913. This figure seemed very low to him.

    Can we offer this student some consolation? Let us assume he is a young man of 25. What is the probability that a man of his age will survive to age 50?

    How is this probability related to the value $l_{50} = 0.913$? What datum is missing for deriving the probability of interest to our student?

    d. Let $l_{25} = 0.964$. What is the probability that a 25-year-old man will reach the age of 50?

2.3.5. ### The Paradox of Chevalier de Méré[12]

A popular historical example from the early days of probability theory concerns seventeenth-century French gamblers who used to bet on the event that in four rolls of a die, at least one ace (i.e., one dot) would turn up. To introduce some variation, they bet on the event that in twenty-four rolls of a pair of dice, at least one double-ace would turn up.

The Chevalier de Méré, a French nobleman of the period, thought that the two events were equally likely. He reasoned that what is true for 4 trials when 6 outcomes are possible, should also be true for 24 trials when 36 outcomes are possible, because $4 : 6 = 24 : 36$. But experience showed the first event to be a bit more likely than the second.

De Méré turned to Blaise Pascal (1623–1662) about this problem, just like the seventeenth-century Italian gamblers who had brought their problem to Galileo (see Problem 2.2.6). Pascal solved the problem with the help of his friend Pierre de Fermat (1601–1665). The two scientists agreed in their correspondence (in 1654) that proportional reasoning, like that of de Méré was inappropriate for inferring the second case from the first. They agreed on a method for computing the two probabilities.

a. Find the probability of obtaining at least one ace in four rolls of a single die.

b. Find the probability of obtaining at least one double-ace in twenty-four rolls of two dice.

2.3.6. A defendant in a paternity suit was given a series of n independent blood tests, each of which *excludes* a wrongfully-accused man with probability p_k, where $1 \leq k \leq n$. If a defendant is not excluded by any of these tests, he is considered a serious

---

[12]See Freedman et al. (1978, pp. 223–225), Freudenthal (1970), Glickman (1989), and Székely (1986, pp. 5–9, including more references).

suspect. If, however, a defendant is excluded by at least one of the tests, he is cleared.

Find the probability, p, that a wrongfully-accused man will in fact be cleared by the series.[13]

2.3.7.   A population is distributed according to the four standard blood types as follows:

$$
\begin{array}{ccc}
\text{A} & - & 42\% \\
\text{O} & - & 33\% \\
\text{B} & - & 18\% \\
\text{AB} & - & 7\%.
\end{array}
$$

Assuming that people choose their mates independent of blood type, calculate the probability that a randomly sampled couple from this population will have the same blood type.

2.3.8.   **How to Get an Answer Without Asking the Question**[14]

A social scientist wanted to find out the percentage of drug users in a population he was studying. Realizing that people are not keen on answering direct questions on the subject, he devised the following procedure:

A big bag was prepared with many question slips. On 70% of the slips the question "Do you use drugs?" was written. On the remaining 30% the question: "Is the last digit of your social security number even?" was written. Each subject randomly drew a slip from the bag, read it, responded to the interviewer with a "yes" or a "no," and destroyed the slip. This way, the interviewer did not know which question the subject answered.

---

[13]Based on an example in Kreith (1976).

[14]Thanks to Ruth Shishinsky for calling our attention to this kind of problem (See Campbell & Joiner, 1973; Chaudhuri & Mukerjee, 1988; Devore, 1979; Fox & Tracy, 1986).

Assume that exactly 50% of the population has social security number with an even last digit, and that all subjects responded truthfully.

   a. It turned out that 44% of the subjects answered "yes." Give an estimate of the proportion of drug users in this population.

   b. What percentage of "yes" answers would have been obtained, had *all* the subjects been using drugs?

2.3.9.   **A Question of Life and Death**

A man condemned to death in the State of Randomana is presented with the following probabilistic situation. He is given two similar jars, 100 white balls, and 100 black balls. He is asked to distribute the balls between the two jars as he wishes. Afterward, he will be blindfolded and allowed to randomly choose one of the jars. Then he will blindly draw one ball from the chosen jar. If he draws a black ball, he will be executed. However, if he draws a white ball, he will be pardoned.

How should this unfortunate prisoner distribute the balls between the jars in the most favorable manner, namely, so that the probability of drawing a white ball at the end of the process is maximal?

2.3.10. Consider the following tables which present admission data of graduate students (by sex) to a university[15] that includes two faculties, F_1 and F_2. Let A be the event 'Applicant is admitted to the university.'

Complete the missing numbers in each table, so as to satisfy the 'inversion' described in the tables, namely, in every faculty women are twice as likely as men to be admitted, yet in

---

[15]Inspired by Bickel, Hammel, and O'Connell (1975), see also Falk and Bar-Hillel (1980).

the university as a whole the probability of a woman being admitted is half that of a man.

In symbols: $P_w(A|F_i) = 2P_m(A|F_i)$, for $i = 1, 2$; and $P_w(A) = 0.5P_m(A)$.

Faculty	Number of applicants	Men - m Admission data (A)	
		Number	Probability
F_1			0.45
F_2			0.05
Total	1000	360	0.360

Faculty	Number of applicants	Women - w Admission data (A)	
		Number	Probability
F_1			0.90
F_2			0.10
Total	1000	180	0.180

See also Problem 2.5.10 (both problems represent the so-called *Simpson's Paradox*. See, e.g., Wagner, 1982).

2.3.11.   The Princess or the Tiger?[16]

Once upon a time (long before the age of feminism) there was a king. The king had a beautiful daughter whom he intended to marry to a prince from a neighboring kingdom. However, a short time before the proposed wedding day, the princess met

---

[16]Based on Armstrong (1982); inspired by a popular short story, "The Lady or the Tiger?" by Frank Stockton which appeared in *A Storyteller's Pack: A Frank R. Stockton Reader*, Scribner, 1968.

Reynaldo — handsome, clever, romantic, but only a peasant — and their love flourished secretly.

Inevitably the king heard to his dismay about their relationship. Irate, he ordered that Reynaldo be thrown into a room full of tigers. But in response to his daughter's pleas, the king offered a compromise: Reynaldo would enter an unfamiliar maze. At every intersection he will have to choose which of the paths to follow. Eventually, he will reach one of two rooms. While the hungry tigers wait in one room, the loving princess waits in the other room. If Reynaldo enters the latter room, he and the princess could marry.

The king showed the princess a map of the maze, like the one below, and let her decide in which room to wait.

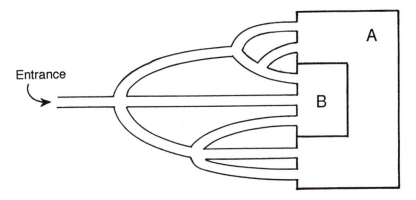

Remember that Reynaldo did not see the map and he can only *guess* which path to follow at each junction.

In which room would you advise the princess to wait? What are Reynaldo's chances of entering each of the rooms?

2.3.12. The Division Paradox[17]

This is another problem that de Méré posed to Pascal and that Pascal resolved (in 1654) with Fermat in their correspondence

---

[17]See Armstrong (1982) and Székely (1986, pp. 9–11).

(see Problem 2.3.5). The problem was first published in the 15th century by Fra Luca Paccioli, but there are indications that it was known much earlier.

Two players play a fair game (i.e., each has a 50% chance of winning). It is agreed that the first player to win 6 rounds is declared the winner and that he takes the stakes. The game gets interrupted after player A has won 5 rounds, and player B has won 3 rounds. How should the stakes be divided fairly?

2.3.13. Assume that the weather can be described by one and only one of two states, *fair* or *rainy*, and that there exists an unequivocal system for determining the state of the weather on any given day. Assume further that the probability that the weather is in a given state depends only on the preceding day's weather.

Let us denote by $P(y \mid x)$ the probability of weather y on a given day, given that the weather on the previous day was x (both x and y may assume the values *fair* or *rainy*).

The weather in a given region may be described by the following conditional probabilities:

$P(rainy \mid fair)=0.2$ $P(rainy \mid rainy)=0.4$
$P(fair \mid fair)=0.8$ $P(fair \mid rainy)=0.6$

It rains Monday. A picnic is planned for Thursday. What is the probability that in three days there will be fair weather?

Hint: Try to draw a 'tree diagram' that describes all the possibilities in a systematic manner.

2.3.14. * **Do Men Have More Sisters Than Women?**

This problem is meant to be addressed to a large class of students.[18]

Suppose each of you were to write down how many brothers and how many sisters you have. We would then collect the notes and separate them into those written by women and those written by men. Should we expect men to have more (or fewer) sisters than women have? Would men have more sisters or more brothers? Or are no differences expected?

2.3.15. Let A and B be two events in a probability space (Ω, P), such that $0 < P(A) < 1$ and $0 < P(B) < 1$. We introduce three definitions concerning the relation between the two events (see Chung, 1942; Falk & Bar–Hillel, 1983; Gudder, 1981; and Problem 7 in Holm, 1970).

(1) B is *independent* of A, denoted $A \perp B$, if knowledge of A's occurrence does not change the probability of B, namely, if $P(B \mid A) = P(B)$.

(2) B is *positively dependent* on A, denoted $A \nearrow B$, if knowledge of A's occurrence increases B's probability, namely, if $P(B \mid A) > P(B)$.

(3) B is *negatively dependent* on A, denoted $A \searrow B$, if knowledge of A's occurrence decreases B's probability, namely, if $P(B \mid A) < P(B)$.

a. Prove that independence is a symmetric relation, that is, if $A \perp B$, then $B \perp A$.

b. Prove that positive dependence is a symmetric relation, that is, if $A \nearrow B$, then $B \nearrow A$.

c. Prove that negative dependence is a symmetric relation, that is, if $A \searrow B$, then $B \searrow A$ (see also Problem 2.3.19).

*A reminder: the asterisk signifies that the answer to the problem includes some discussion or extension.

[18]See Falk (1982), Falk and Konold (1992), and Mosteller (1980).

2.3.16. Prove that if A and B are independent events, so are $\overline{A}$ and $\overline{B}$.

2.3.17. Using the notation of Problem 2.3.15, show that:

   a. If $A \nearrow B$, then $A \searrow \overline{B}$.
   b. If $A \nearrow B$, then $\overline{A} \searrow B$.

2.3.18. An urn contains w white balls and b black balls ($w \geq 1$ and $b \geq 1$). The balls are thoroughly mixed, and two are drawn, one after the other, *without* replacement. Let W_i and B_i denote the respective outcomes 'white on the ith draw' and 'black on the ith draw,' for $i = 1, 2$.

Prove that $P(W_2) = \frac{w}{w+b}$, as is the case for $P(W_1)$.

This clearly implies that $P(B_2) = P(B_1) = \frac{b}{w+b}$.

2.3.19. * An urn contains two white balls and two black balls. We blindly draw two balls, one after the other, *without* replacement from that urn.

W_1 will denote the outcome 'a white ball on the first draw.'

W_2 will denote the outcome 'a white ball on the second draw.'

   a. Find $P(W_2 \mid W_1)$.
   b. Find $P(W_1 \mid W_2)$.

2.3.20. During a class-party at the end of a probability course, a prize was offered for guessing the precise outcomes when two dice, one white and one black, were tossed.

One student believed she could increase her probability of winning so that it was higher than 1/36. Her strategy, which follows, was particularly debated:

It is known that the sums obtained when throwing two dice are not all equally likely. The possible sums (from 2 to 12) are triangularly distributed, with 7 as the modal result. Therefore, one should offer a guess in which the sum is 7, namely, randomly pick an ordered pair of dice-outcomes from those with sum 7.

What is the winning probability when employing this strategy? (Inspired by Paulson, 1992)

2.3.21. **How to Change The Sex Ratio**

A story about a (male) monarch tells that he wished to change the sex distribution of the population in favor of women, so that each man would be able to have a big harem.[19] Unlike ruthless rulers of the past, he chose 'humane' methods for that end. He ruled that every woman could continue to bear babies as long as they were females. However, if a woman gave birth to a boy, she had to stop having babies.

If that regulation was followed, then only certain family types would exist in that population. These are (according to birth order): M, FM, FFM, FFFM, ... Would such a policy, if carried out long enough, fulfill the ruler's expectations? Could it change the sex ratio in favor of males? Or would it fail to change the approximately equal proportions of males and females after all?

---

[19]We thank Tibor Nemetz for telling us that story (in Budapest, in 1981). See Konold (in press).

2.4　Bayes' Theorem

2.4.1.　A doctor is called to see a sick child. The doctor knows (prior to the visit) that 90% of the sick children in that neighborhood are sick with the flu, denoted F, while 10% are sick with the measles, denoted M. Let us assume for simplicity's sake that M and F are complementary events.

A well-known symptom of measles is a rash, denoted R. The probability of having a rash for a child sick with the measles is 0.95. However, occasionally children with the flu also develop a rash, with conditional probability 0.08.

Upon examining the child, the doctor finds a rash. What is the probability that the child has the measles? (Denote the required probability accurately, using the above symbols.)

2.4.2.　A man was arrested on suspicion of murder. Let us denote the event 'the man is guilty' by G. The investigating officer collected all the relevant information, added his impressions of the suspect, and arrived at the conclusion that the man's probability of guilt was 0.60.

    a. As the investigation went on, it was learned (beyond any doubt) that the murderer's blood type was O. The relative frequency of blood type O in the population is 0.33 (i.e., this is the probability that a randomly selected person in the population has blood type O). The suspect's blood was tested and found to be O.

    Compute the posterior probability of this suspect's guilt (from the officer's point of view) considering all the data.

    b. Suppose both the murderer's and the suspect's blood types were found (with certainty) to be A. The relative frequency of blood type A in the population is 0.42. How would the posterior probability of guilt in that case compare with the same probability in **a**? Would it be greater, smaller, or equal? Explain.

2.4.3. Consider the case of a suspect in a criminal case. Let G denote the event that the suspect is guilty. $P(G)$ is the suspect's prior probability of guilt. Let E denote evidence obtained in the subsequent inquest.

In each of sections **a-d** you are asked to give the posterior probability of guilt, $P(G \mid E)$, based on the given information. Computations are hardly necessary.

    a. $P(G) = 0.60$; $\quad P(E \mid G) = P(E \mid \overline{G}) \neq 0$.

    b. $P(G) = 0.75$; $\quad P(E \mid G) \neq 0$; $\quad P(E \mid \overline{G}) = 0$.

    c. $P(G) = 0.60$; $\quad \dfrac{P(E \mid G)}{P(E \mid \overline{G})} = 2$.

    d. $P(G) = 0.50$; $\quad P(E \mid G) = 1$.

2.4.4. **Probabilistic Battleship**[20]

Consider a modified version of the game 'battleship', which students often play in class when bored. There are two players. Each has a grid with 3 cells A, B, and C, and each hides a battleship in one of his or her three cells. The players alternate "shooting" at one of the other player's cells.

A	B	C
?	?	?

Unlike the original game, a shot that hits a cell containing a ship does not necessarily sink the ship; it only sinks it with

---

[20]Based on Hering (1987).

probability p (where $0 < p < 1$), and with probability $1 - p$ the shooter is informed "No," signifying a failure to sink the ship. The shooter is also informed "No" if he or she shoots an empty cell. Obviously, the game ends when one ship sinks.

Three conditional probabilities thus describe the rules of the game:

$$P(\text{game ends} \mid \text{ship is in cell}) = p;$$

$$P(\text{"No"} \mid \text{ship is in cell}) = 1 - p;$$

$$P(\text{"No"} \mid \text{cell is empty}) = 1.$$

a. Let $p = 1/2$. The game can be conducted so that somebody who is not playing at the moment *secretly* flips a coin whenever the cell containing the ship is hit. If the outcome is H the ship is sunk and the game ends. If the outcome is T a "No" is announced and the game goes on.

Before starting the game, your probabilities that the ship is in cell A, B, or C of your opponent's board are equal (see line (0) of the table on the next page).

(1) Suppose you shoot cell A and you get the response "No." Fill in line (1) of the table with the revised probabilities in light of your new information.

(2) You now have to make another shot. Obviously, you want to choose the cell with maximum probability of containing the ship. Let's agree that if more than one cell has a maximal probability you will select the

Probability of Containing the Ship

		Cell		
	S t a g e	A	B	C
(0)	prior to any shooting	1/3	1/3	1/3
(1)	following "No" to cell A			
(2)	following "No" to cell __			
(3)	following "No" to cell __			

letter which is first alphabetically. Fill in the name of the cell you shoot in line (2) of the table.

Suppose that shot also receives a "No" response. Fill in the newly-revised probabilities in line (2).

(3) Now at what cell should you shoot? Fill in your choice in line (3). Suppose you again receive a "No" response. Fill in the updated probabilities in line (3).

b. Same problem as **a**, but now $p = 2/3$. Make a similar table for that case.

2.4.5. * Assume that people can be sorted unequivocally into two distinct sets according to their hair color: dark — denoted D, and blond — denoted B.

A person's hair color is determined by two alleles of a gene, each transmitted *at random* by one parent. The allele for dark hair, denoted d, is dominant over that for blond hair, denoted b. Hence, of the three genotypes, dd, db, and bb the first

---

* A reminder: the asterisk signifies that the answer to the problem includes some discussion or extension.

two would result phenotypically in D, only the third would be B. One can be certain that a blond person is homozygous (i.e., bb). However, upon observing a dark-haired person, one cannot know whether that person is genotypically homozygous (dd) or heterozygous (bd).

Consider a couple with both mates dark haired and heterozygous for hair color. The genotypes and phenotypes of the potential offspring of the couple are given in the table below:

Mother D

		d	b
Father D	d	dd D	db D
	b	bd D	bb B

The probability of such a couple giving birth to a dark-haired child is 3/4 and to a blond child is 1/4. The probability of a random D child of such parents being heterozygous is 2/3.

A couple is interested in knowing whether any of their future children could be blond.[21] Both husband and wife are dark haired. Their phenotypes, along with those of their parents and siblings, are shown in the figure on the next page. Note that each of these prospective parents has two dark-haired parents and a blond brother.

Let H denote the genotypic event, or the hypothesis that the couple has the *potential* to produce a blond child. H is equal to the intersection of the events that husband and wife are heterozygotes (i.e., each is a carrier of a recessive gene b). $\overline{H}$

---

[21]See Falk (1983).

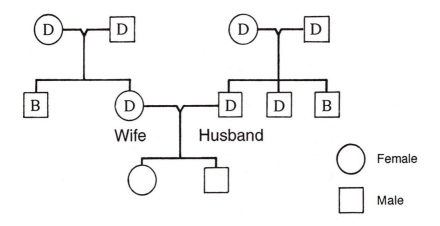

is the event that the couple is incapable of producing a blond child (i.e., at least one spouse is homozygous dd). If H is true, then the probability of that couple having a B child is $1/4$.

    a. Find the prior probability of H (before any children are born in that family). Denote this $P_o(H)$.

    b. A dark-haired baby is born to the couple. Denote this event D_1 (first-born child dark). Has the probability of H changed? Denote the probability of H after the birth of the first child by $P_1(H)$. What is the value of $P_1(H)$?

    c. After a few years the couple has another baby, which is *blond*. Denote this event B_2. What is the posterior probability of H in light of this information? In line with the previous notations, what is $P_2(H)$?

2.4.6.   In a help center[22] for rape victims the following information is given to women: statistics show that most rape victims were assaulted at home, and only a small minority were assaulted while hitch-hiking. Ergo, the outcry against hitch-hiking is

---

[22]Suggested by Dorit Rivkin.

exaggerated. Hitch-hiking is not that dangerous. Home is more dangerous.

Express your opinion about these conclusions (we do not question the reliability of the data). Are they justified? Explain.

2.4.7. It is known that the risk of a newborn baby having Down's Syndrome is strongly linked to the mother's age.[23] For a 30-year-old woman the probability of giving birth to such a baby is 1/885. For a 45-year-old woman it is 1/32.

Amniocentesis is known to be a very accurate test. The test indicates 99.5% of the normal embryos as such and mistakenly diagnoses Down's Syndrome in only 0.5% of the normal embryos. The rate of error is similar for the affected embryos: 99.5% of them are identified as such, and only 0.5% of them are mistakenly diagnosed as normal.

Let us denote the event 'embryo affected by Down's Syndrome' by 'S' and the event that the test result is positive (embryo diagnosed as affected) by '+.' A negative test result (embryo diagnosed as unaffected) will be denoted by '−.'

A pregnant woman is subjected to amniocentesis, and a positive answer is obtained. This answer worries the woman very much, especially considering the reputation of the test as highly accurate. Accordingly, she weighs her next steps.

   a. Denote by symbols, based on those given above, the probability of most interest to the woman.

   b. Should a positive result worry a 30-year-old woman as much as a 45-year-old one? Compute the probability (of **a**) for a woman of each of the two ages, and base your answer on the calculated results.

---

[23]Following Pauker and Pauker (1979).

2.4.8. Psychologists at a mental-health center use a clinical test to diagnose psychotic patients. The test correctly identifies 80% of all tested psychotics ('valid positives'). On the other hand, it also wrongly diagnoses 15% of the healthy individuals as psychotics ('false positives').

It is known that 10% of those who come to the center are psychotics.[24]

    a. What is the probability that testing a random person arriving at the center will yield a correct diagnosis?

    b. What is the probability that the test will diagnose a random testee as psychotic?

    c. When the test indicates that a person is psychotic, what is the probability that he/she is in fact psychotic?

    d. Discuss the efficacy of the test by comparing the relevant probabilities.

2.4.9.   **A Problem with Two Liars**[25]

It is known that Tom and Dick tell the truth only a third of the time (the probability that each is lying is 2/3, every time).

Tom makes a statement, and Dick tells us that Tom was speaking the truth.

What is the probability that Tom was actually telling the truth?

2.4.10.   **The Three Liars**[26]

Tom, Dick, and Harry are three liars. Each of them tells the truth only a third of the time (the probability that each is lying is 2/3, every time).

---

[24]Based on an example of Meehl (1956, p. 265).

[25]Inspired by Eddington (1935). See also Falk (1986b), O'Beirne (1961), and Zabell (1988b).

[26]See footnote to Problem 2.4.9.

Tom makes a statement, and Harry tells us that Dick said that Tom was speaking the truth.

What is the probability that Tom was actually telling the truth? (Do not get entangled in symbols. Derivation of the required probability is not immediate. Try a 'tree' to reach the likelihoods that should be inserted into the Bayesian calculation.)

2.4.11.   * **How Many Sons?**[27]

Mrs. F. is known to be the mother of two. You meet her in town with a boy whom she introduces as her son. What is the probability that Mrs. F. has two sons? One possible answer is that the probability is 1/2: you now have seen one boy, and hence the question is whether the other child is a male. Another possibility is that the probability is 1/3: you learned that Mrs. F. has 'at least one boy,' and hence three equiprobable family structures are possible (BB, BG, GB), of which the target event (BB) is but one.

Which is the correct answer? Explain.

(Hint: What are your assumptions about the chance mechanism that yielded your observation of Mrs. F. and her son?)

2.4.12.   * **Doctor Fischer's Bomb Party**[28]

Graham Greene's Dr. Fischer is intrigued by the question "How greedy are the rich?" He invites six wealthy guests to a party. When they are all gathered he shows his guests a barrel of bran in a corner of the garden. There are six Christmas crackers in the barrel. Five of the crackers, he explains,

---

[27]See Bar-Hillel and Falk (1982), Chu and Chu (1992), Falk and Konold (1992), and Glickman (1982).

[28]Based on an episode in Graham Greene's (1980) novel. Inspired by Ayton and McClelland (1987).

contain cheques for two million Swiss Francs. The sixth contains enough explosive to end the life of the person pulling the cracker. The guests are invited to approach the barrel one-by-one and try their luck.

While one of the guests stands up and pauses to gather courage to approach the barrel, Mrs. Montgomery beats him by running to the barrel first. She has figured that "the odds would never be as favorable again" (Greene, 1980, p. 127).

Although everybody knows the cheques are in the crackers, the issue is complicated by a mention of the possibility that the presence of a bomb is just a hoax.

Suppose the chances are 50/50 that the bomb is there.

How justified was Mrs. Montgomery in pushing ahead of everybody to pull the first cracker? Compute the probabilities (before the onset of the 'game') of the first, second, ..., sixth player blowing up. Which (if any) is the safest serial position? (Compare with Problems 2.3.3 and 2.4.13).

2.4.13.   * **Seek and You Shall Find:**
The Hope Function throughout a Search Process[29]

Imagine that you are searching for an important letter that you received some time ago. Usually your assistant puts your letters in the drawers of your desk after you have read them. He remembers to do this in 80% of the cases, and in 20% of the cases he leaves them somewhere else.

There are eight drawers in your desk. If indeed your assistant *has* placed the letter in your desk, you know from past experience that it is *equally likely* to be in any of the eight drawers.

You start a thorough and systematic search of your desk.

---

[29]Developed, with Abigail Lipson, on the basis of a problem by Meshalkin (1963/1973, p. 21).

a. (1) You search the first drawer, and the letter *is not* there. What is *now* the probability that the letter is *in the desk?*

(2) You continue to search the next three drawers, until altogether you have searched four drawers. The letter *is not* there. What is the probability *now* that the letter is *in the desk?*

(3) You continue to search three more drawers, until altogether you have searched seven drawers. The letter *is not* there. What is the probability *now* that the letter is *in the desk?*

b. Everything is as before, except for a different question (make sure to notice the difference).

(1) You search the first drawer, and the letter *is not* there. What is *now* the probability that the letter is *in the next drawer* (i.e., in the second drawer)?

(2) You continue to search the next three drawers, until you have searched four drawers. The letter *is not* there. What is the probability *now* that the letter is *in the next drawer* (i.e., in the fifth drawer)?

(3) You continue to search three more drawers, until you have searched seven drawers. The letter *is not* there. What is the probability *now* that the letter is *in the next drawer* (i.e., in the eighth drawer)?

Compare the situation described in this problem with that of Problem 2.4.12.

2.4.14.   *** The Car or the Goat?**

In a famous TV game show, "Let's Make a Deal," that has recently been the focus of much discussion,[30] the contestant is

---

[30]The number of publications on this problem is growing exponentially these days. The following is by no means an exhaustive list: repeated discussions in the "Ask Marilyn" column of *Parade Magazine* (September 9, 1990, p. 15; December

given a choice of three closed doors. Behind one of the doors is an attractive prize (e.g., a car); behind the other two are gag prizes (e.g., goats). Suppose the contestant picks door No. 1. It remains closed and the host, Monty, who knows what's behind the doors, opens one of the other doors, say No. 3, to reveal a goat.[31] He then gives the contestant the option of switching to No. 2. Should the contestant stick or switch?

2.5 Probability Distributions and Expectations

2.5.1. In a research study on animal behavior, mice are given a choice between four similar doors. One of them is the 'correct' door. If a mouse chooses the correct door it is rewarded with food, and the experiment ends. If it chooses an incorrect door the

---

2, 1990, p. 25; February 17, 1991, p. 12; July 7, 1991, p. 29), "Behind Monty Hall's Doors: Puzzle, Debate and Answer?" by John Tierney *The New York Times*, July 21, 1991 and letters to the editor in August 11, 1991, Falk (1992), Gardner (1992), Gillman (1991, 1992), Morgan, Chaganty, Dahiya and Doviak (1991), Saunders (1990), Selvin (1975a, 1975b), Shaughnessy and Dick (1991).

[31]It should be clear that, by the rules of the game, the host will always open a door hiding a goat.

mouse is punished with a mild electric shock and then brought back to the starting point to choose again.

Let X denote the number of trials in the experiment, that is, the ordinal number of the trial on which the first correct choice occurs. Find the probability distribution of X in each of the three cases below:

  a. All the doors are equally likely to be chosen on each trial, independently of the events on previous trials (a dumb mouse).

  b. On each trial, the mouse chooses, with equal probabilities, one of the doors that has not been chosen up to that moment. A door that has been tried is never chosen again (an intelligent mouse).

  c. The mouse never chooses the same door on two consecutive trials, otherwise it chooses among the available doors with equal probabilities (an intelligent mouse with very short memory).

2.5.2. An enthusiastic sport fan decides to express his support for his team by betting on it, and eventually making some profit out of that enjoyable activity.

He knows from past statistics that the probability of his team winning, in a given match, is 0.75, and the probability of losing is 0.25 (let us ignore other possible outcomes).

  a. He gets \$12 if his team wins. How much should he be willing to pay, if his team loses, in order to make an average profit of \$2 per bet?

  b. He pays \$15 beforehand, and if his team wins he receives \$20. What is his expected gain from this bet?

2.5.3. **All the Eggs in One Basket**

A cook bakes a cake containing eight eggs. She breaks the eggs into a bowl one after the other. If a rotten egg enters the bowl it is thrown away together with all the eggs previously placed in the bowl, and the cook stops baking the cake. Each egg costs \$0.09. The probability of any egg being rotten is 0.05.

What is the expected loss using this method? (There is no loss if all the eight eggs are wholesome and the cake is baked.)

2.5.4. An urn contains 7 balls: 3 red and 4 blue. Balls are randomly drawn from the urn, one after the other, *without* replacement.

Let X be the number of red balls drawn before drawing the first blue ball.

    a. Construct the probability-distribution table for X.

    b. What is the expected value of X?

Compare these two questions with the corresponding questions in Problem 2.5.5.

2.5.5. Consider an urn comprising 3 red balls and 4 blue balls, as in Problem 2.5.4.

Balls are randomly drawn from that urn, one after the other, *with* replacement.

Let X be the number of red balls drawn before drawing the first blue ball.

    a. Give the formula for the probability distribution of X.

    b. What is the expected value of X?

Compare with the corresponding questions in Problem 2.5.4.

2.5.6. A man returns home at night during a blackout (without a flashlight) with a bunch of n keys, one of which will unlock his front door. He tries to unlock the door with one key after the other, each chosen *at random* from the bunch at hand, independently of previous choices. Find his expected number of trials until the door is unlocked. (See Feller, 1957, pp. 46 & 54).

    a. If he removes from the bunch every key that is tried and found unsuitable.

    b. If keys that are found unsuitable are *not* removed from the bunch and can be chosen and tried again (the man is probably drunk).

2.5.7.   **Chuck-a-Luck**

The game of "chuck-a-luck" is played at carnivals in the Midwest and England (see Gardner, 1982, pp. 102–103; Mosteller, 1965, pp. 20–21; Paulos, 1988, p. 36). Its rules are as follows: you pick a number from 1 to 6 and the operator rolls three dice. If the number you picked comes up on all three dice, the operator pays you $3; if it comes up on two dice, he pays you $2; and if it comes up on just one die, he pays you $1. Only if the number you picked doesn't come up at all do you pay him anything – just $1. Indeed, it seems tempting.

    a. What is the probability that you'll win money playing this game?

    b. What is your expected gain (in dollars) from playing the game?

    c. If you are still paid $2 if your number comes up twice and $1 if it comes up once, how much money should you be offered if your number comes up on all three dice for your expected gain to be exactly zero?

2.5.8. A family about to fly overseas for a long stay decides to send ahead their luggage by sea. The value of the property to be sent is $2500. They consider insuring their luggage. An insurance agent has offered them a deal involving a premium of $150 and securing a compensation of $3000 in case of total loss of their property.

The probability that the luggage will not reach its destination (will burn, drown, be stolen, etc.), according to extensive statistics of past experience, is 0.01. Ignoring the delivery costs (and the possibility of partial loss), compute the family's expected 'gain' from the transaction of sending their luggage by sea:

    a. without any insurance.

    b. if they buy the above insurance.

Suppose the family buys the insurance. Would you consider that decision 'rational'?

2.5.9. A medical clinic tests blood for a certain disease from which approximately one person in a hundred suffers.[32] People come to the clinic in groups of 50. The director wonders whether he can increase the efficiency of the testing procedure by conducting pooled tests.

Suppose that, instead of testing each individually, he would pool the 50 blood samples and test them all together. If the pooled test was negative, he could pronounce the whole group healthy. If not, he could then test each person's blood individually.[33]

What is the expected number of tests the director will have to perform if he pools the blood samples?

---

[32]Based on an example by Paulos (1988, pp. 35–36).

[33]Ballinger and Benzer (1989) apply a more complex modification of the same pooling principle in a molecular-biological context.

2.5.10. The most popular measure which demographers use to characterize mortality in a population in a given year is the Crude Death Rate (CDR) — defined as the ratio of the number of deaths during the year to the average size of the population. The CDR is thus the probability that a random person from the population studied will die during the year.

The Age-Specific Death Rate (ASDR) is the ratio of the number of deaths (during a given year) in a given age group to the average size of the population of that age. Thus, for every age x, the specific death rate is the probability that a random person of age x will die within a year.

For the last several decades in Israel, the following demographic pattern has been consistently observed: The CDR of Jews was higher than that of Arabs. On the other hand, in each single age group, the ASDR for Arabs was higher than for Jews. Hence, an investigator who wishes to compare the mortality of the two subpopulations will reach one conclusion when relying on the CDR and an opposite conclusion when relying on the ASDRs (Falk & Bar-Hillel, 1980).

Explain how such an apparently-paradoxical situation could occur.

2.5.11. Suppose you draw a random sample of size n out of a set of N elements, N_1 of which are of a specified category. Drawing an element of that category is considered a *success*.

    a. Find the probability distribution of the random variable X, defined as the number of successes in the sample when the sampling is conducted *with replacement*.

    b. Find the probability distribution of the random variable Y, defined as the number of successes in the sample when the sampling is conducted *without replacement*.

    c. Try to present the formulas you found in **a** and in **b** so that both their similarity and differences are visible. The similarity in their structures is due to their similar verbal descriptions; the difference is due to their methods of sampling.

2.5.12. * Find the expectations of the binomial and the hypergeometric random variables, as (respectively) defined in parts **a** and **b** of Problem 2.5.11. (See Falk, 1986c).

2.5.13. * **How Many Coincidences Are Expected?**

    a. A bag contains notes bearing the four numbers 1, 2, 3, and 4. A player is invited to blindly draw them, one by one, and place them, in turn, in boxes numbered 1, 2, 3, 4. The player will be paid one dollar for every number placed in the 'correct' box.

    Compute the player's expected winnings.[34]

    b. Let Y be the number of number-notes placed in the correct box. Construct the probability-distribution table of Y, and check the result obtained in **a** by computing $E(Y)$.

    c. Suppose the situation described in **a** refers to n number-notes and n boxes. Find the expected winnings. How does the expected number of matches (i.e., notes in identically-numbered boxes) depend on n?

2.5.14. Consider a random binary sequence of length N comprised of X's and O's. Let N_1 and N_2 respectively be the number of X's and O's in the sequence (both N_1 and N_2 are at least one; clearly, $N_1 + N_2 = N$).

---

* A reminder: the asterisk signifies that the answer to the problem includes some discussion or extension.

[34]Based on Freudenthal (1970, pp. 161–162).

Let R denote the *number of runs* in the sequence. For example, in the sequence below (where $N_1 = 7$ and $N_2 = 6$) the random variable R assumes the value 5 (as depicted):

$$\underline{O\ O}\ \underline{X\ X\ X\ X\ X\ X}\ \underline{O}\ \underline{X}\ \underline{O\ O\ O}.$$

Find a formula for the probability distribution of R (it is helpful to consider the cases of an even and an odd number of runs separately).

2.5.15. Consider a random binary sequence of N symbols, as in Problem 2.5.14. The number of runs in such a sequence is denoted R. Your task is to compute $E(R)$.

A direct computation of $E(R)$, based on the probability distribution presented in the answer to Problem 2.5.14, is not easy. An indirect approach, using the method of adding expectations of elementary random variables, proves much easier (Falk, 1986c).

Since there are N symbols in the sequence, there are $N - 1$ *transitions* from one symbol to the next. On each transition one can have either an alternation, that is, change of symbols, or a continuity. Consider the random variable A defined as the *number of alternations in a sequence*. In the example in Problem 2.5.14, $A = 4$ (i.e., there are four changes of symbols out of 12 transitions). The maximal value A can attain is $N - 1$, and the minimum is 1.

Clearly, $A = R - 1$ (indeed, $R = 5$ in the sequence in Problem 2.5.14). Therefore, $E(R) = E(A) + 1$. If we know $E(A)$, we know $E(R)$. Try to find a way to decompose the random variable A into a sum of more elementary random variables, and then obtain $E(A)$ by the addition law of expectations.

2.5.16. Consider the following games of chance:

Game 1. You roll a fair die 50 times. The number of points you get equals the sum of the 50 outcomes.

Game 2. You roll a fair die once. The number of points you get equals the outcome times 50.

    a. Answer the questions concerning the *number of points gained*. Fill in the entries of the table below:

	Game 1	Game 2
Shape of the distribution		
Expected value		
Variance		
Standard deviation		

    b. What is the probability of gaining 300 points in Game 1? In Game 2?

2.5.17.   * **Where the Grass Is Greener**[35]

You are shown the back side of two cards. On the front of one of them is written a positive number, and on the other, half that number. One of the cards is randomly selected and shown to you. You may either win the number of dollars shown on this card or switch cards and win the number of dollars shown on the other card. What should you do?

The following argument shows that you should always switch to the other card:

---

[35] Adapted from the "Flaws, Fallacies, and Flimflam" column of the *College Mathematics Journal* (January, 1990, Vol. **21**, p. 35; suggested by R. Guy). See also Falk and Konold (1992) and Zabell (1988a).

Suppose the number revealed to you is A. Then either $2A$ or $0.5A$ is written on the other card, each with probability $1/2$. If you stick with the original card your (expected) winning is A. If you switch, your expected winning is $(1/2)2A + (1/2)0.5A = 1.25A$. Thus, you should always prefer the card other than the one revealed to you.

Moreover, even without having seen the number on the first card, you should switch, since the same reasoning applies to *any positive* A. But then you should also switch back, and so on. You are paradoxically caught in a never-ending pendulum swing. What is wrong here?

Chapter 3

Reasoning across Domains

3.1 Statistics, Probability, and Inference

3.1.1. Test A was given to a large class. The distribution of scores, denoted x, is characterized by several measures. The names of these measures (and their symbols) are written in the first column of the table on the next page. The numerical values of these measures are written in the second column.

The teachers' council concluded that the test had been too 'harsh' and considered two schemes for adjusting the scores of all students.

  a. Five points would be added to each student's original score.

  b. Each score would be increased by 15% of its original value (i.e., multiplied by 1.15).

Fill the table with the measures characterizing the adjusted distribution of scores according to each of the two schemes.

Measures characterizing three distributions of scores

Measure	original scores x	adjusted scores scheme **a** $x+5$	adjusted scores scheme **b** $1.15x$
Linear correlation coefficient with scores in test B (i.e., r)	0.55		
Rank-order correlation coefficient with scores in test C (i.e., r_s)	0.61		
Arithmetic mean (M)	58		
Standard deviation (σ)	12		
Variance (σ^2)	144		
Median (Me)	59		
Standard score of student with the lowest achievement $(z_{\min})$	-2.8		
Range (R)	64		
Mode (Mo)	60		
Percentile of student i (i.e., F_i)	78%		

3.1.2. A statistician performed a squaring transformation on a set of n nonnegative x values, and obtained a set of y values such that $y = x^2$.

The following is a list of measures characterizing the distribution of the variable x. For each measure, check whether squaring the x-measure gives the corresponding y measure. That is, given that m is some measure of the distribution, determine whether $m_x^2 = m_y$.

    a. $\overline{x}$.

    b. σ_x .

    c. $Me(x)$.

    d. $Mo(x)$.

    e. $G(x)$, the geometric mean.

    f. $Max(x)$.

    g. $Q_3(x) - Q_1(x)$, the interquartile range.

    h. $R(x)$, the range.

    i. $r_{x,u}$, Pearson's correlation coefficient with variable u.

    j. $r_s(x, v)$, Spearman's rank-order correlation coefficient with variable v.

    k. $z(\overline{x})$.

    l. $\sigma_{z(x)}$.

    m. $F(x_j)$.

3.1.3. A statistics teacher told four students their scores in a somewhat playful manner. He informed them that the distribution of the class's scores was approximately normal with mean 70 and standard deviation 9, and he gave them the following information about their own achievements:

Jack: "Your score splits the class. One quarter has higher scores; three quarters have lower scores."

Jill: "Your raw score is 88."

John: "Your standard score is -1.4."

Jane: "Only 10% of the class got higher scores than yours."

Fill in the missing scores in the table on the next page.

Student's name	Raw score	Standard score	Percentile
Jack			75
Jill	88		
John		-1.4	
Jane			90

3.1.4.   A population's mean is 60, and its standard deviation is 5.

In each case below, a random variable is defined and information about the population's distribution is given. Find the probability that each random variable will either be greater than 66 or less than 54.

    a. The random variable is a single random observation from the population (Note that nothing is known about the population's distribution).

    b. The population's distribution is *normal.* The random variable is a single random observation from the population.

    c. The population's distribution is *normal.* The random variable is the *mean* of a *random sample* of 4 observations from the population.

    d. The population's distribution is *normal.* The random variable is the *mean* of a *random sample* of 18 observations from the population.

Note. Make sure you understand the reason for the direction of change of your answers, from **a** to **d**.

3.1.5. A normally-distributed population has an unknown mean μ_x and a standard deviation $\sigma_x = 20$.

    a. What is the probability that a random observation X from this population will fall within the range $\mu_x \pm 2$ (i.e., will satisfy $\mu_x - 2 \leq X \leq \mu_x + 2$)?

    b. What is the probability that the mean of a random sample of size 144 from the population will fall within the range $\mu_x \pm 2$?

    c. What is the probability that the mean of a random sample of size 400 from the population will fall within the range $\mu_x \pm 2$?

Note. Make sure you understand the reason for the direction of change of your answers, from **a** to **c**.

3.1.6. Insert the missing equality or inequality sign (one and only one of the symbols $=$, $<$, or $>$) between the following pairs of expressions.

    a. We refer to two sequences of length n (where $n > 2$), one composed of x's and the other of y's, such that there is positive variance within each sequence and the two sequences are partly but not fully correlated (i.e., $0 <| r_{x,y} |< 1$).
Let $\hat{y}$ denote the y values predicted by the regression line of y on x, and let $\hat{x}$ denote the x values predicted by the regression line of x on y.

    (1) $z(\overline{x})$ $\overline{z(x)}$

    (2) $\overline{x^2}$ $(\overline{x})^2$

    (3) $|\sum_{i=1}^{n}(x_i - \overline{x})|$ $\sum_{i=1}^{n}| x_i - \overline{x} |$

(4) n $\displaystyle\sum_{i=1}^{n} z_i^2$

(5) $3\sigma^2(x)$ $\sigma^2(3x)$

(6) $r_{(3x+5),(5y-2)}$ $r_{x,y}$

(7) $\sigma_x \sigma_y$ $cov(x,y)$

(8) $\displaystyle\sum_{i=1}^{n}(y_i - \overline{y})^2$ $\displaystyle\sum_{i=1}^{n}(y_i - \widehat{y}_i)^2$

(9) $\dfrac{|\,x_i - \overline{x}\,|}{\sigma_x}$ $\dfrac{|\,\widehat{y}_i - \overline{y}\,|}{\sigma_y}$

(10) $\dfrac{\sigma_y}{\sigma_x}$ $b_{y.x}$

(11) $\sigma_{\widehat{x}}^2$ σ_x^2

(12) $\overline{y}$ $\overline{\widehat{y}}$

(13) $b_{y.x} b_{x.y}$ $|\,r_{x,y}\,|$

(14) $P(|\,z_x\,| \geq k\sigma_x)$ $\dfrac{1}{k^2 - \epsilon}$

where $k > 0$ and ϵ is very small $(\epsilon < k^2)$ and positive.

b. Let $\overline{x}_n$ denote the mean of a random sample of n values: $x_1, x_2, \ldots, x_n$ (for $n > 2$), from a population (with a finite variance σ_x^2) whose mean is μ_x.

(1) $\mu\left(\dfrac{\displaystyle\sum_{i=1}^{n}(x_i - \overline{x}_n)^2}{n}\right)$ σ_x^2

$$(2) \qquad \mu(\overline{x}_{65}) \qquad\qquad \mu(\overline{x}_{22})$$

$$(3) \qquad \sigma^2(\overline{x}_{65}) \qquad\qquad \sigma^2(\overline{x}_{22})$$

$$(4)\ P(|\ \overline{x}_{10} - \mu_x\ | > 2.7) \qquad\qquad P(|\ \overline{x}_{100} - \mu_x\ | > 2.7)$$

3.1.7. The table below describes two populations and some statistics based on random samples drawn from these populations. Answer, in the appropriate cells, the questions concerning the sampling distributions of the statistics (M_n denotes the mean of a sample of size n).

	Population			
	Expectation 80; variance 16; shape of distribution unknown		Normal distribution of x; expectation: $\mu_x = 120$; variance (σ_x^2) unknown	
Statistic	M_{250}	$\dfrac{M_{400} - 80}{\sqrt{16/400}}$	$\dfrac{M_{18} - 120}{\sqrt{\dfrac{\displaystyle\sum_{i=1}^{18}(x_i - M_{18})^2}{17 \times 18}}}$	Number of observations exceeding 120 in a sample of size 12: $N(x \mid x > 120)$
Shape of sampling distribution				
Expectation of sampling distribution				
Variance of sampling distribution				

3.1.8. The random variable X is normally distributed, with $\mu_x = 80$; $\sigma_x^2 = 81$. An independent variable Y is also normally distributed, with $\mu_y = 100$; $\sigma_y^2 = 100$.

Five random variables are defined in the rows of the table on the next page. Answer the questions concerning the distributions of these variables, as described at the heads of the columns. Write your answers in the appropriate cells.

Let $\overline{U}_n$ denote the mean of n random observations of the variable U. The X and the Y values are sampled independently of each other.

Random Variable	Questions about the distribution		
	Shape	Expectation	Variance
$\overline{X}_{27} + \overline{Y}_{25}$			
$2 - 3\overline{X}_{36}$			
$\dfrac{(\overline{Y}_{10} - \overline{X}_9) - 20}{\sqrt{\frac{100}{10} + \frac{81}{9}}}$			
$\left(\dfrac{X - 80}{9}\right)^2 + \left(\dfrac{Y - 100}{10}\right)^2$			
The proportion of Y values exceeding 108.4 in a random sample of size 500: $\dfrac{N(Y \mid Y > 108.4)}{500}$			

3.1.9. You wish to estimate the median of a population.[1] Eight random values are independently sampled from the population. Remember: The probability of obtaining a value below the median is equal to that of obtaining a value above the median (ignore the possibility of equality).

---

[1]See Noether (1980).

a. You decide to use the interval between the lowest and the highest of the eight sample values to estimate the median.

Compute the probability that this confidence interval will indeed include the median of the population.

Hint: Compute via the probability of the complementary event.

b. Suppose you decide to use the interval between the fourth largest value in the sample (of 8 values) and the next one (the fifth largest value in the sample) to estimate the population median. Compute the probability that this confidence interval will include the population median.

3.1.10. The experiment described below is designed to clarify the difference between 'classical' inference (significance testing) and Bayesian inference (Falk, 1986a).

Suppose you confront 10 opaque urns. You know that all of them contain 7 beads, and the urns are divided into two types:

Nine urns of type A, each containing 5 white beads and 2 black beads.
One urn of type B, containing 5 black beads and 2 white beads.

You randomly choose one of the ten urns and formulate two complementary hypotheses about it:

H_0: The urn is of type A.

H_1: The urn is of type B.

You now apply a decision procedure (in line with classical statistical inference). You blindly draw two beads (without replacement) from the urn in question. If the two beads are black, H_0 is rejected and H_1 is accepted, otherwise, H_0 cannot be rejected.

Let R denote the event that H_0 is *rejected* (i.e., two black beads are drawn).

> **a.** Compute the level of significance of the test, that is, $\alpha = P(R \mid H_0)$.
>
> **b.** Compute the Bayesian posterior probability of H_0, given R, that is, $P(H_0 \mid R)$.

3.1.11. * Two cab companies, the blues and the greens, operate in a given town; 85% of the cabs in town are blue, 15% are green.

One night a cab was involved in a 'hit and run' accident. An eyewitness claimed that the cab was *green*. The court examined the discrimination capacity of the witness, given the illumination conditions at the time and site of the accident. They determined that the witness identified colors correctly in 80% of the cases presented to him (and made a wrong judgment in 20% of the cases).

> **a.** What is the probability that the cab involved in the accident was indeed green?[2]
>
> You have no information whatsoever about the driving records of the two companies, nor about their differential distributions in various parts of the town. You may therefore start by regarding the cab involved in the accident as if it were *randomly* picked from the totality of cabs in town.
>
> **b.** Upon further investigations, another eyewitness was found. The new witness *independently* supported the first witness's testimony, namely, that the car involved was a green cab. It turned out that this witness's capacity for color discrimination was equal to that of the first witness.
>
> Two judges tried the case:
>
> Judge C (the 'classic') decided that he would find the green cab company guilty if the probability of getting

---

* A reminder: the asterisk signifies that the answer to the problem includes some discussion or extension.

[2]Until here the problem is based on Tversky and Kahneman (1980).

two such testimonies, given that the hit-and-run cab was blue, is less than 5%.

Judge B (the 'Bayesian') decided to calculate the posterior probability that the hit-and-run cab was green, given the two testimonies. He decided to find the green company guilty if that probability exceeded 95%.

Carry out the calculation suggested by each judge and state the conclusion that follows from it. Which verdict would you support? Explain.

3.1.12. A Society for Beyond the Sensory announced the opening of a new journal, and invited investigators in the behavioral sciences to submit experimental papers in the field. The editor let know that he will publish any paper that would present results *significant* at a level of 0.05, indicating extra-sensory perception (ESP).

Ten investigators responded to the challenge and, independently of each other, started experiments designed to find out whether their subjects perceive extra-sensory messages above random (guessing) level.

Assume that none of the subjects perceive anything and that they just *guess*. What is the probability that *at least one* paper will be published (as a result of these 10 experiments), heralding significant ESP findings?

3.1.13. **On Replications**

There is a wide consensus among scientists and methodologists that replications are crucial for establishing the validity of research results. Some scientists are extreme in their insistence on the central role of replications in scientific investigation: Guttman (1977) declares that "The essence of science is replication" (p. 86), and Carver (1978) maintains that "Replication is the cornerstone of science .... If results are due to chance, then results will not replicate" (p. 392).

Even fairy-tale wisdom acknowledges the merits of replication, as expressed in the following verse translated from the Hebrew version of Rumpelstiltskin. This conversation takes place after the miller's daughter has managed (with the dwarf's help) to spin straw into gold for the first time:

Treasurer:	My lord the king, consider, then,
	The maiden further to detain.
King:	What for? What should we gain?
Treasurer:	The purpose, lord, is very clear:
	Again to test her, while she's here.
King:	Another test — that's your intent?
Treasurer:	It could be chance, an accident ...
Chamberlain:	Blind luck, a fluke, some happenstance —
	A miracle occurred perchance?
Treasurer:	If Lady Fortune's hand was here,
	A second time she'll not appear.
	But if 'tis true as we are told,
	We'll know that straw is really gold!

> From "Utz Li Gutz Li" Avraham Shlonski
> (1966, p. 60; translated by Oren Falk)

Let us examine quantitatively some of the contingencies involving two independent replications of a given experiment.

Two investigators conduct the same kind of experiment, on two independent random samples from the same population. Both test the same null hypothesis.

a. Suppose H_0 is true, and each of the tests is conducted at a level of significance of 0.05.

Compute the following:

(1) The probability that both tests will turn out statistically significant.

(2) The probability that at least one experimenter will obtain a statistically-significant result.

(3) The probability that only one of the tests will come out statistically significant.

(4) The probability that neither test will turn out statistically significant.

(5) The conditional probability that the second investigator will obtain a statistically-significant result, given that the first one did so.

b. Suppose H_1 is true, and the power of each of the tests is 0.80.

Compute the same five probabilities as in **a**.

3.1.14. * **Probability, Correlation, and Inbreeding**[3]

The genetic concept of *inbreeding* is used when mates, who are more closely related than they would be if they had been chosen at random from the population, produce offspring. Related mates have one or more ancestors in common. Consequently, the two alleles that an inbred individual (i.e., an offspring of related mates) inherits, for a particular gene, may both originate from the same allele in a common ancestor. Inbreeding thus increases the probability that an individual is homozygous above its expected value when genes combine independently.

For a given gene there are two separate processes which can cause an individual to be homozygous. The two alleles may be (1) identical by descent, in that both are derived from the same allele in a common ancestor, and (2) alike in state, that is to say, indistinguishable by their biological function, but

---

[3]The problem and its solution have been worked out in collaboration with Arnold Well (See Falk & Well, in preparation).

randomly paired. The *inbreeding coefficient, I*, is defined as the *probability that two paired alleles originated from the same allele in a common ancestor* (Crow & Kimura, 1970, pp. 64–66; Roughgarden, 1979, pp. 181–184).

Consider a gene with two possible alleles, whose probabilities in the population are p and q (clearly, $p + q = 1$). We may assign these alleles the values 1 and 0, respectively. Given the probability I (as defined above), we can compute the probability that an individual will be of any given genotype. For example, there are two ways both alleles can have the value 1: if they are derived from the same allele of the same ancestor (with probability I) and have the value 1 (with probability p), or if they are randomly combined (with probability $1 - I$) and both happen to have value 1 (probability $p \cdot p$).

a. The table below is constructed to present the joint probability distribution of the allele values received from each parent, for the case of two alleles and inbreeding coefficient I, $0 \leq I \leq 1$. Write the missing probabilities in the empty cells.

Probabilities of All Possible Genotypes, with Inbreeding Coefficient I: Two Alleles

Value of sperm: Y	Value of egg: X		
	0	1	Total
1			p
0			q
Total	q	p	1

b. An alternative way to quantify inbreeding is to address the relatedness between an individual's parents. A straightforward index of relatedness is the correlation between the genetic contribution of the two mates (Crow & Kimura, 1970, pp. 68–69).

Let r represent the correlation between the genetic values of two gametes that unite to produce an individual.

Prove that $r = I$, where I is the inbreeding coefficient for that individual. (Base your answer on the joint probability distribution of X and Y, as computed in **a**).

c. To find out how general the equality is between I and r, let us explore the converse situation. We start with an arbitrary 2×2 joint probability distribution (of variables X and Y) whose correlation coefficient is r. Can we justifiably claim that r conveys the probability of "identity by descent"? (or the "fraction by which heterozygosity is reduced"? See answers to parts **a** and **b** of this problem). For the question to be meaningful, we have to limit our consideration to variables X and Y with equal marginal distributions and assume that the marginal probabilities, p and q, are given. This constraint implies that the two-dimensional distribution is symmetric about the secondary diagonal. The table below presents such a distribution:

Y	X		Total
	0	1	
1	$p_{01} = p_{10}$	p_{11}	p
0	p_{00}	p_{10}	q
Total	q	p	1

The equal probabilities in the two cells of the principal diagonal reflect the distribution's symmetry.

Show that the joint probabilities, p_{ij}, of such a distribution are uniquely determined by the given r value and the marginal probability p. Furthermore, show that the probabilities are determined in such a way that r can be interpreted as an inbreeding coefficient (with p and q representing the population proportions of the two alleles).

d. Consider a gene with n possible alleles. Let us assign any set of n values, $a_1, a_2, \ldots, a_n$, to these alleles, and denote their respective probabilities in the population by $p_1, p_2, \ldots, p_n$ (where $\sum_{i=1}^{n} p_i = 1$). If A denotes the value assigned to a random allele of the gene in the population, then $p_i = P(A = a_i)$ for $i = 1, 2, \ldots, n$.

Let p_{ij} denote the probability that an individual receives a pair of alleles of states i and j. Let X denote the value of the egg allele, and Y the value of the sperm allele. Then $p_{ij} = P[(X = a_i) \cap (Y = a_j)]$ for $i, j = 1, 2, \ldots, n$. Clearly, $p_{ij} = p_{ji}$, and the marginal distributions (of X and Y) are equal.

If I is the inbreeding coefficient, namely, the probability that an individual's pair of alleles are identical by descent, then $1 - I$ is the probability that the two alleles are independently paired. The probabilities of all the possible genotypic combinations can be obtained by the same method as in the case of two alleles (see **a**). Find a formula for the joint $n \times n$ probability distribution of X and Y. Let us call this distribution a *genetic distribution*.

e. Prove that in the general case of n alleles, I is equal to r, that is, the correlation coefficient between the values of the uniting gametes (as has been shown for the case of two alleles — see **b**). In fact, you have to show that the correlation coefficient of a genetic distribution equals the inbreeding coefficient.

Part II

MULTIPLE-CHOICE PROBLEMS

Chapter 4

Descriptive Statistics II

Choose the most correct answer (one and only one) for each problem.

4.1 Scales of Measurement [1]

4.1.1. When measurements are performed in an *interval* scale, which of the following is *not permitted*?

   a. Calculating the difference between numbers.

   b. Calculating the ratio of differences between numbers.

   c. Calculating the ratio between numbers.

   d. Changing the zero point.

   e. Changing the unit of measurement by multiplying all numbers by a positive constant.

   f. Performing a positive linear transformation on the numbers.

---

[1]A good introduction to the subject of scales of measurement, including additional references, can be found in Siegel and Castellan (1988, pp. 23–33).

4.1.2.   When measurements are performed in an *ordinal* scale, which of the following *is allowed?*

    a. Multiplying the numbers by a positive constant.

    b. Calculating the difference between numbers.

    c. Exchanging all the numbers with some other set of numbers, provided there is one-to-one correspondence between the numbers in the two sets.

    d. Calculating the ratio between numbers.

    e. Calculating the ratio of the differences between numbers.

    f. Replacing all the numbers by their absolute values.

4.1.3.   In which measurement scale *can't* one subtract a constant value from each of the numbers?

    a. Nominal scale.

    b. Ordinal scale.

    c. Interval scale.

    d. Ratio scale.

    e. One can do it in all scales.

4.1.4. In which measurement scale can one replace all the measurements by a set of new numbers — provided there is a one-to-one correspondence between the new set and the original one — without loss of information?

    a. In all the scales.

    b. In none of the scales.

    c. In an absolute scale.

    d. In a ratio scale.

    e. In an interval scale.

    f. In an ordinal scale.

    g. In a nomimal scale.

4.1.5. Several objects were measured and assigned positive numbers. A statistician suggested taking the logarithm of each number. Which measurement scale would allow such a transformation without losing information?

    a. Ordinal scale.

    b. Interval scale.

    c. Ratio scale.

    d. All scales allow it.

    e. None of the scales allow it.

4.1.6.   A teacher prepared a lesson on measurement scales[2] and numbered them, according to the level of measurement, in the following way: 1 – nominal; 2 – ordinal; 3 – interval; 4 – ratio; 5 – absolute.

By assigning numbers to the scales the teacher in effect measured their level. Which measurement scale did the teacher use?

> **a.** Nominal.
>
> **b.** Ordinal.
>
> **c.** Interval.
>
> **d.** Ratio.
>
> **e.** Absolute.

4.1.7.   You are given a set of objects of different weights. You have a balance with two pans, but no standard weights. You use the balance to weigh all the objects and assign numbers to their weights. What is the highest level of measurement you can achieve?

> **a.** Nominal.
>
> **b.** Ordinal.
>
> **c.** Interval.
>
> **d.** Ratio.
>
> **e.** Absolute.

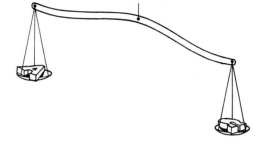

---

[2]Inspired by a comment in Eisenbach (1988, p. 10).

4.1.8. The employees of a certain plant are assigned eight ranks labeled (from the lowest to the highest) 1, 2, ..., 8. Employees of rank 1 receive a monthly salary of $2000; those of rank 2 receive $2400, etc.; for each additional rank employees are paid $400 more a month. If we regard the eight ranks as measures of the level of payment the employee receives, what is the level of this measurement scale?

    a. Nominal.

    b. Ordinal.

    c. Interval.

    d. Ratio.

    e. Impossible to tell on the basis of this information.

4.2 Measures Characterizing Distributions II

4.2.1. A marathon runner decides to spare his energy for the end of a race and to also not lag too far behind the other runners in the earlier parts of the race. For a while he adjusts his speed so that half the runners overtake him and half lag behind him.

Which average of the racers' speeds is represented by the above runner's speed?

    a. The mode.

    b. The arithmetic mean.

    c. The midrange.

    d. The geometric mean.

    e. The median.

4.2.2.   The mean score of six students is 10. Five students got the following scores:

1, 7, 11, 13, 17.

What is the sixth student's score?

  a. 16

  b. 15

  c. 12

  d. 11

  e. 10

  f. 9

  g. 7

  h. 6

4.2.3.   Which of the averages that characterize a set of numbers is *necessarily* equal to some number(s) in the set?

  a. All of the averages.

  b. None of the averages.

  c. The mode.

  d. The arithmetic mean.

  e. The median.

  f. The midrange.

  g. The geometric mean.

4.2.4. You are given n pairs of numbers: x_1, y_1; x_2, y_2; $\ldots$; x_n, y_n. The symbol $\overline{xy}$ is shorthand for which of the following computations?

    a. $\dfrac{x_1 + x_2 + \ldots + x_n}{n} \cdot \dfrac{y_1 + y_2 + \ldots + y_n}{n}$.

    b. $\dfrac{x_1 x_2 \ldots x_n}{n} + \dfrac{y_1 y_2 \ldots y_n}{n}$.

    c. $\dfrac{x_1 y_1 + x_2 y_2 + \ldots + x_n y_n}{n}$.

    d. $\dfrac{x_1 + y_1}{2} \cdot \dfrac{x_2 + y_2}{2} \cdot \ldots \cdot \dfrac{x_n + y_n}{2}$.

    e. $\dfrac{(x_1 + y_1)(x_2 + y_2) \ldots (x_n + y_n)}{n}$.

4.2.5. Let x and y be two numbers. We will use the expression xMy to signify $(x + y)/2$. The symbol M thus represents the *operation* of computing the arithmetic mean of two numbers. Which of the following claims is always true?

    a. $aMb = bMa$ and $(aMb)Mc = aM(bMc)$, that is, the operation M is commutative and associative.

    b. $aMb = bMa$ and $(aMb)Mc \neq aM(bMc)$, that is, the operation M is commutative and is not associative.

    c. $aMb \neq bMa$ and $(aMb)Mc = aM(bMc)$, that is, the operation M is not commutative and it is associative.

    d. $aMb \neq bMa$ and $(aMb)Mc \neq aM(bMc)$, that is, the operation M is not commutative and not associative.

(See Hadar & Hadass, 1981a; Walter, 1981)

4.2.6.   Let A denote the arithmetic mean of a set of positive numbers, not all of which are equal, and let G denote the geometric mean of the same numbers. Which of the following is true?

   a. $G < A$

   b. $A < G$

   c. $A = G$

   d. Impossible to know. It dependes on the particular numbers.

(See Problem 1.1.19 in Part I)

4.2.7.   The number 29 is added to each of the biggest 200 numbers in a set of 401 numbers.

   How will this addition affect the median of the set of numbers?

   a. One cannot tell. It depends on the particular numbers in the set.

   b. The median will increase by 29.

   c. The median will increase by 200.

   d. The median will increase by 29×200.

   e. The median will increase by $\frac{29}{401}$.

   f. The median will increase by $\frac{200 \times 29}{401}$.

   g. The median will not change.

4.2.8. There were n students in a class ($n \geq 4$). Some of the students appealed their scores on a certain test. The papers were reviewed, and the scores of four students were raised from 70 to 90. How does this change affect the mean score of the class?

> **a.** One cannot know. It dependes on all the students' exact scores.
>
> **b.** The mean does not change.
>
> **c.** The mean increases by 20.
>
> **d.** The mean increases by $\frac{20}{n}$.
>
> **e.** The mean increases by 80.
>
> **f.** The mean increases by $\frac{80}{n}$.
>
> **g.** The mean increases by $\frac{20}{4}$.

4.2.9. In a group of 30 girls, the median score in verbal ability is 80. For the same variable, but in a group of 20 boys, the median score is 70. Which of the following claims concerning Me, the median score for the combined group of 50 boys and girls, is the most accurate and correct?

> **a.** Nothing can be known about the value of Me because the 50 scores are missing.
>
> **b.** $Me = 75$.
>
> **c.** $Me = \dfrac{30 \times 80 + 20 \times 70}{50} = 76$.
>
> **d.** $Me = 150$.
>
> **e.** $70 \leq Me \leq 80$.

4.2.10. The median age of a class of students was 23 when starting their university studies and 28 at graduation (everybody
graduated eventually). Which of the following statements,
concerning $Me(D) = $ *the median duration of studies* at the
university, is necessarily true?

  a. $Me(D) < 5.$

  b. $Me(D) > 5.$

  c. $Me(D) = 5.$

  d. There is insufficient data to draw any of the conclusions
  in **a**, **b**, or **c**.

4.2.11. What is the length of the interquartile range, $Q_3 - Q_1$, in all
the distributions?

  a. $\dfrac{R}{2}.$

  b. $\dfrac{Me}{2}.$

  c. $2 \times 0.67 \times \sigma.$

  d. 50%.

  e. 25%.

  f. Impossible to know; it depends on the specific distribution.

4.2.12. A teacher checks a list of mathematics scores (x scores) of her class. She wishes to summarize the data. Which measure could she use to express the extent of individual differences in achievement between her students?

    a. The mean of standard scores.

    b. The mode.

    c. $\overline{x^2}$.

    d. The upper quartile.

    e. The standard deviation.

    f. $(\overline{x} - Me)/\sigma_x$.

4.2.13. Which of the following formulas defines a measure of *dispersion* among the observations x_1, x_2, x_3, ..., x_n?

    a. $\dfrac{\sum\limits_{i=1}^{n}(x_i - \overline{x})}{n}$.

    d. $\dfrac{\sum\limits_{i=1}^{n}(x_i - \overline{x})^2}{n}$.

    b. $\sqrt{\dfrac{\sum\limits_{i=1}^{n}x_i^2}{n}}$.

    e. $\dfrac{x_{min} + x_{max}}{2}$.

    c. $\left(\dfrac{\sum\limits_{i=1}^{n}x_i}{n}\right)^2$.

    f. $\dfrac{\overline{x} - Me}{\sigma_x}$.

4.2.14. What is the standard deviation of the following set of numbers: 10, 10, 4, 4?

    a. $\sqrt{3}$

    b. $\sqrt{6}$

    c. 3

    d. 6

    e. 9

    f. 12

    g. 36

4.2.15. In which of the following sets of numbers does the variance equal 6?

    a. 5, 8, 11.

    b. 4, 16.

    c. 6, 12, 18.

    d. 3, 9.

4.2.16. Two students from the same class dropped out of school. These students' scores in literature were equal to the class's mean.

How does their leaving affect the measures characterizing the distribution of the class's literature scores?

    a. The mean decreases and the variance decreases.

    b. The mean decreases and the variance does not change.

    c. The mean does not change and the variance does not change.

    d. The mean increases and the variance increases.

    e. The mean does not change and the variance increases.

    f. The mean does not change and the variance decreases.

4.2.17. The mean salary in a certain plant was $1500, and the standard deviation was $400. A year later each employee got a $100 raise. After another year each employee's salary (including the above mentioned raise) was increased by 20%. What are the mean and the standard deviation[3] of the current salaries in dollars?

    a. Mean 1920, standard deviation 400.

    b. Mean 1900, standard deviation 400.

    c. Mean 1920, standard deviation 416.

    d. Mean 1900, standard deviation 600.

    e. Mean 1920, standard deviation 600.

    f. Mean 1900, standard deviation 580.

    g. Mean 1920, standard deviation 480.

    h. Mean 1900, standard deviation 480.

---

[3] Adapted from a problem composed by Yaacov German

4.2.18. A community includes k families. Let x_i denote the number of children in the ith family ($1 \le i \le k$). What is the mean number of siblings of a child in this community?[4]

a. $\dfrac{\sum\limits_{i=1}^{k} x_i}{k}$.

b. $\dfrac{\sum\limits_{i=1}^{k} (x_i - 1)}{k}$.

c. $\dfrac{\sum\limits_{i=1}^{k} x_i^2}{k}$.

d. $\dfrac{\sum\limits_{i=1}^{k} x_i(x_i - 1)}{\sum\limits_{i=1}^{k} x_i}$.

e. $\dfrac{\sum\limits_{i=1}^{k} (x_i - 1)^2}{\sum\limits_{i=1}^{k} x_i}$.

f. $\dfrac{\sum\limits_{i=1}^{k} (x_i - 1)^2}{k^2}$.

g. $\dfrac{\sum\limits_{i=1}^{k} x_i^2}{\sum\limits_{i=1}^{k} x_i}$.

h. $\dfrac{\sum\limits_{i=1}^{k} x_i(x_i - 1)}{k}$.

---

[4]Inspired by Yaacov German. Compare with Problem 1.1.18 of Part I.

4.3 Transformed Scores, Association, and Linear Regression

4.3.1. The list of test results for students in a certain course was published in standard scores. Jill saw the number 1.5 next to her name. She decided to compute this number's standard score[5] in relation to the list of numbers published. What was the result of her computation?

    a. 1.5

    b. 1.0

    c. -1.5

    d. 0

    e. 0.5

    f. One cannot know because information on the mean and standard deviation of the scores is missing.

4.3.2. A set of scores, x_1, x_2, x_3, $\ldots$, x_n (that are not all the same) was translated into a set of standard scores, z_1, z_2, z_3, $\ldots$, z_n. Which of the following claims about the set of standard scores *is not necessarily true*?

    a. The sum of the positive standard scores is equal to the sum of the absolute values of the negative standard scores.

    b. The number of positive standard scores is equal to the number of negative standard scores.

    c. The mean of the standard scores is zero.

    d. The variance of the standard scores is 1.

    e. The sum of the squared standard scores is n.

---

[5]Suggested by Yaacov German.

4.3.3.   All the math scores of 26 children in one class were converted
to standard scores, denoted as follows:

$$z_1, \ z_2, \ z_3, \ldots, \ z_{26} \ .$$

The data below concern the distribution of the above standard
scores. Which of the following is *impossible?*

   a. $\sum_{i=1}^{26} z_i^2 = 30.$

   b. The maximal standard score is $z_{max} = 3.3.$

   c. $\sum_{i=1}^{26} z_i = 0.$

   d. 15 of the standard scores are negative.

   e. The median standard score is $-0.3.$

4.3.4.   In a certain school it was decided to transform the scores of
students in each class to standard scores on the background
of the class's scores. John studies in a class of 36 students.
He got a standard score 2 in English. What conclusion can be
drawn from these data?

   a. John is the best student in English in his class.

   b. John got the median English score in his class.

   c. The standard deviation of scores in English in the class
   is 2.

   d. John is the next best student in English in his class.

   e. John is the second in an increasing order of English scores
   in his class.

   f. John is among the 10 best students in English in his class.

4.3.5. A researcher analyzed a set of students' scores. She transformed each score into a *standard score* (in relation to the total set) and into a *percentile* (i.e., the percent of scores less than or equal to the student's score).

Which of the following is necessarily true?

    a. There is no association whatsoever between the standard scores and the percentiles.

    b. A zero standard score corresponds to the 50th percentile.

    c. The linear correlation coefficient (Pearson's) between the standard scores and the percentiles is 1.

    d. The rank-order correlation coefficient (Spearman's) between the standard scores and the percentiles is 1.

4.3.6. A class of 30 students took two tests. Let x denote the score on the first test, and let y denote the score on the second test. It turned out that $\bar{x} = \bar{y}$ and $\sigma_x = \sigma_y$. What can we conclude about the size of the correlation coefficient, $r_{x,y}$, between the scores of these two tests?

    a. One cannot conclude anything about $r_{x,y}$ from these data.

    b. The correlation is complete (either $+1$ or -1).

    c. $r_{x,y} > 0$.

    d. $r_{x,y} = 0$.

    e. $r_{x,y} > 1/2$.

4.3.7.   Medical researchers report a *negative correlation* between body weight and frequency of osteoporosis (loss of bone density) among older women.

What is the meaning of this finding?

    a. The chances of diagnosing osteoporosis are greater among heavier women (than among low-weight women).

    b. The chances of diagnosing osteoporosis are greater among lighter women (than among heavy women).

    c. Overweight causes osteoporosis in women.

    d. Underweight causes osteoporosis in women.

    e. Osteoporosis makes women lose weight.

    f. There is no association between body weight and osteoporosis in women.

4.3.8.   The following is a list of pairs of x and y measurements of seven people. Let $r_{x,y}$ denote the linear correlation coefficient (Pearson's) based on these numbers, and let r_s denote the rank-order correlation coefficient (Spearman's) based on the same data. Which of the statements on the next page is true with respect to both coefficients? Answer *without* calculating.

x	y
7	62
3	56
15	95
6	57
8	70
10	73
9	72

a. $r_s = 1$ and $r_{x,y} = 1$.

b. $r_s = 1$ and $0 < r_{x,y} < 1$.

c. $0 < r_s < 1$ and $0 < r_{x,y} < 1$.

d. $0 < r_s < 1$ and $r_{x,y} = 0$.

e. $r_s = 0$ and $r_{x,y} = 0$.

f. $0 < r_s < 1$ and $r_{x,y} = 1$.

4.3.9. Whenever the functional relation between x and y is as described in the figure below:

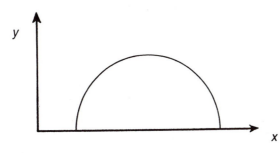

a. x and y are correlated and dependent.

b. x and y are correlated and independent.

c. x and y are noncorrelated and dependent.

d. x and y are noncorrelated and independent.

4.3.10. Pearson's correlation coefficient between x and $\hat{y}$ (i.e., the y value predicted by x via the regression line) is $r_{x,\hat{y}} = 1$. What can be inferred[6] about the correlation between x and y?

> **a.** Nothing can be inferred from the given information concerning $r_{x,y}$.
>
> **b.** $r_{x,y} = 1$.
>
> **c.** $r_{x,y} = 0$.
>
> **d.** $r_{x,y} = -1$.
>
> **e.** $-1 \le r_{x,y} < 0$.
>
> **f.** $0 < r_{x,y} \le 1$.

4.3.11. You know that $r_{x,y} = -0.75$. Denote by $\hat{y}$ the y-value predicted by the regression line (of y on x).

> (1) $r_{x,\hat{y}}$ equals:
> > **a.** -0.75.
> >
> > **b.** 1.
> >
> > **c.** 0.75.
> >
> > **d.** -1.
> >
> > **e.** $0.75^2 = 0.5625$.
> >
> > **f.** Impossible to know for lack of sufficient data.
>
> (2) $r_{y,\hat{y}}$ equals:
> > **a.** -0.75.
> >
> > **b.** 1.
> >
> > **c.** 0.75.
> >
> > **d.** -1.
> >
> > **e.** $0.75^2 = 0.5625$.
> >
> > **f.** Impossible to know for lack of sufficient data.

---

[6]Suggested by Yaacov German.

4.3.12. (1) Complete the missing value in the table below so that the φ coefficient will be *negative*.

	x	
y	0	1
1	10	?
0	20	40

    a. 40

    b. 30

    c. 20

    d. 10

(2) Complete the missing value in the table below so that the φ coefficient will be *zero*. [7]

	x	
y	0	1
1	10	?
0	5	2

    a. 20

    b. 10

    c. 5

    d. 4

    e. 2

    f. 1

---

[7]We thank Yaacov German for suggesting this problem.

4.3.13. Each of the following four figures depicts two regression lines. The symbol $\hat{y}$ denotes the regression line of y as a function of x, and $\hat{x}$ denotes the regression line of x as a function of y. Which of these diagrams represents an *impossible* situation?

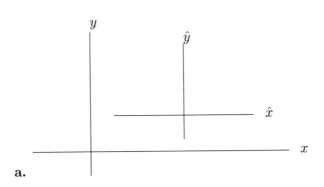

a.

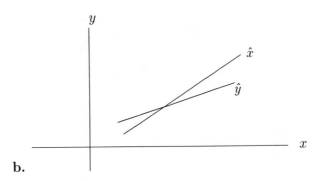

b.

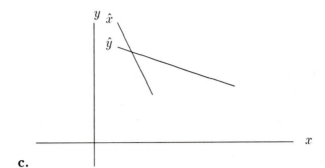

c.

d.

4.3.14. The phenomenon of *regression to the mean* in prediction of y by x is strongest when:

    a. $b_{y \cdot x} = 1$.

    b. $\mid r_{x,y} \mid = 1$.

    c. $r_{x,y} = 1/2$.

    d. $r_{x,y} = 0$.

    e. $b_{y \cdot x} = b_{x \cdot y} \neq 0$.

4.4   In Retrospect

4.4.1.   Two measurements, x and y, were taken of each of n objects. Which of the following measures is *not* a pure number, and therefore must be accompanied by appropriate units?

a. $Z_{Me(x)}$ (the median of the x's — standardized).

b. R_x (the range of the x's).

c. $r_{x,y}$ (Pearson's correlation coefficient between x and y).

d. $\dfrac{\bar{y} - Me(y)}{\sigma_y}$ (measure of skewness of the y's).

e. $\dfrac{(Q_3(x) - Me(x)) - (Me(x) - Q_1(x))}{Q_3(x) - Q_1(x)}$ (measure of skewness of the x's).

f. $V(x) = \dfrac{\sigma_x}{\bar{x}} 100$ (coefficient of variation of the x's).

g. $r_s(x, y)$ (Spearman's rank-order correlation coefficient between x and y).

Chapter 5

Probability II

Choose the most correct answer (one and only one) for each problem.

5.1.1. At the London meteorological center the daily weather forecasts were recorded over a long period. The days in which the weatherman predicted that the probability of rain the next day was 80% were recorded. The daily weather was also written down, including whether it rained or not on each day.

How can one validate the weatherman's accuracy[1] on the days in which he assessed the chance of rain the next day as 80%?

    a. He can be considered accurate if it rained on all the days for which the probability 80% was given.

    b. He can be considered accurate if it rained on 80% of the days for which the probability 80% was given.

    c. One cannot determine whether the weatherman was accurate or not because the phenomenon is uncertain.

    d. He can be considered accurate if it rained on half of the days for which that probability was given, and it didn't rain on half of them.

    e. He can be considered accurate if it rained on most of the days for which that probability was given.

---

[1]See Konold (1989) and Lichtenstein, Fischhoff, and Phillips (1982).

111

5.1.2. *A* and *B* are two events in a probability space. We know only that $P(A) = 0.30$ and $P(B) = 0.40$. Which of the following is *always* true?

    a. $P(A \cup B) = 0.35$.

    b. $0.30 \le P(A \cup B) \le 0.40$.

    c. $0.40 \le P(A \cup B) \le 0.70$.

    d. $P(A \cup B) \le 0.30$.

    e. $P(A \cup B) \ge 0.70$.

    f. $P(A \cup B) = 1$.

    g. $P(A \cup B) = 0.58$.

5.1.3. *A* and *B* are two events in a probability space. $P(A) = 0.25$ and $P(B) = 0.60$. Which of the following is *always* true?

    a. $P(A \cap B) = 0.15$.

    b. $0.60 \le P(A \cap B) \le 0.85$.

    c. $0.25 \le P(A \cap B) \le 0.60$.

    d. $P(A \cap B) = 0.425$.

    e. $P(A \cap B) \le 0.25$.

    f. $P(A \cap B) = 0$.

    g. $P(A \cap B) = 0.70$.

5.1.4. If $P(A \cup B) = P(A)$, then

    a. $A = \Omega$.

    b. $B = \phi$.

    c. $A = B$.

    d. $B \subseteq A$.

    e. $A \subseteq B$.

5.1.5. If $P(A \cap B) = P(A)$, then

    a. $B = \Omega$.

    b. $A = \phi$.

    c. $A = B$.

    d. $B \subseteq A$.

    e. $A \subseteq B$.

5.1.6. Which of the following relations between $P(B \mid A)$ and $P(A \cap B)$ is true for any two events A and B, provided that $P(A) \neq 0$?

    a. $P(A \cap B) \leq P(B \mid A)$.

    b. $P(A \cap B) \geq P(B \mid A)$.

    c. $P(A \cap B) = P(B \mid A)$.

    d. One cannot tell; the relation depends on the specific definitions of the events A and B.

5.1.7.   Let A and B be two *disjoint* events, with $0 < P(A) < 1$ and $0 < P(B) < 1$. Which of the following is *always* true?

    a. $P(A \mid B) = P(B)$.

    b. A and B are independent events.

    c. A and B are dependent events.

    d. $P(A \cup B) = 0$.

    e. $P(A) + P(B) = 1$.

5.1.8.   If a child is infected with a given contagious disease, the probability that another child will become infected upon contact with the sick child is 0.7.

John is sick with this disease. What is the probability that the third child who contacts John will be the first one who is infected by him?

    a. 0.343

    b. 0.147

    c. 0.027

    d. 0.973

    e. 0.657

    f. 0.063

5.1.9.   Let A and B be two *independent* events in a probability space, such that $0 < P(A) < 1$ and $0 < P(B) < 1$. Let $A \triangledown B$ denote the symmetric difference of A and B, that is, the event of the occurrence of *either A or B* but *not both* (see answer to Problem 2.1.8a in Part I). Which of the following is true?

    a. $P(A \triangledown B) = P(A) + P(B)$.

    b. $P(A \triangledown B) = P(A) + P(B) - P(A)P(B)$.

    c. $P(A \triangledown B) = 1 - P(A)P(B)$.

    d. $P(A \triangledown B) < P(A \cup B)$.

    e. $P(A \triangledown B) > P(A \cup B)$.

5.1.10. Let A be an event in a probability space (Ω, P), such that $0 < P(A) < 1$. Which of the following is *incorrect*?[2]

    a. $P(\Omega \mid A) = 1$.

    b. $P(A \mid \Omega) = P(A)$.

    c. $P(\overline{A} \mid A) = 0$.

    d. $P(A \mid A) = P(A)$.

---

[2]Suggested by Yaacov German.

5.1.11. You know that two events A and B in a probability space satisfy the following: $A \subset B$, $P(A) > 0$, and $P(B) < 1$.

Which of the following is *not* necessarily true?

    a. $P(\overline{B}) \leq P(\overline{A})$.

    b. $P(A \mid B) = 1$.

    c. A and B are dependent events.

    d. $P(B - A) = P(B) - P(A)$.

    e. $P(B \mid A) = 1$.

5.1.12. If A, B, and C are three events such that $P(A)P(B)P(C) > 0$, and $P(B) > P(A)$, which of the following is *always* true?

    a. $A \subset B$.

    b. $P(\overline{A}) > P(\overline{B})$.

    c. $P(\overline{A}) > P(B)$.

    d. $P(A) > P(\overline{B})$.

    e. $P(\overline{B}) > P(\overline{A})$.

    f. $P(B \mid C) > P(A \mid C)$.

    g. $P(C \mid B) > P(C \mid A)$.

    h. $P(B \cap C) > P(A \cap C)$.

5.1.13. A multiple–choice test consists of three problems. For each problem there are four choices, only one of which is correct. A student comes totally unprepared and decides to answer by sheer guessing. What is the probability that he'll answer at least one problem correctly? [3]

a. $\dfrac{3}{4}$ **e.** $\dfrac{37}{64}$

b. $\dfrac{2}{3}$ **f.** $\dfrac{27}{64}$

c. $\dfrac{9}{16}$ **g.** $\dfrac{7}{16}$

d. $\dfrac{1}{2}$

5.1.14. A mechanical system is composed of k components, all of which must work in order for the system to function. The probability that each component works is p. What is the probability that the system will *fail* to function?

     a. p^k.

     b. $(1-p)^k$.

     c. $1 - kp$.

     d. $1 - k(1-p)$.

     e. $1 - p^k$.

     f. $1 - (1-p)^k$.

---

[3]Suggested by Yaacov German.

5.1.15. It is given that $P(A \mid B) = 0.3$ and $P(A \mid \overline{B}) = 0.7$.

Which of the following claims, regarding the probability of the event A, is necessarily true?

    a. One cannot tell a thing about the value of $P(A)$.

    b. $0 \le P(A) \le 0.3$.

    c. $0.3 \le P(A) \le 0.7$.

    d. $0.7 \le P(A) \le 1$.

    e. $P(A) = 0.5$.

    f. $P(A) = 1$.

5.1.16. A and B are two events in a probability space. You know that $0 < P(A) < 1$.

Which of the following is always true?

    a. $P(B \mid A) + P(B \mid \overline{A}) = 1$.

    b. $P(B \mid A) + P(B \mid \overline{A}) = P(B)$.

    c. $P(B \mid A) + P(B \mid \overline{A}) = P(A)$.

    d. $P(B \mid A) + P(B \mid \overline{A}) = P(A \cap B)$.

    e. $P(B \mid A) + P(B \mid \overline{A}) = P(A \cup B)$.

    f. None of these claims are true.

5.1.17. Let A and B be two events such that $0 < P(A) < 1$, $0 < P(B) < 1$, and $P(B \mid A) > P(B)$. Which of the following must be true? (see Problem 2.3.15 in Part I.)

   a. $P(A \mid B) < P(A)$.

   b. $P(\overline{B} \mid A) > P(\overline{B})$.

   c. $P(A \mid B) > P(A)$.

   d. $P(A \cap B) < P(A)P(B)$.

   e. $P(A \mid B) = P(A)$.

5.1.18. Let A and B be two events satisfying $0 < P(A) < 1$, $0 < P(B) < 1$, and $P(B \mid A) = P(B \mid \overline{A})$. Which of the following must be true?

   a. $P(A \mid B) = P(\overline{A} \mid B)$.

   b. $P(B \mid A) = P(\overline{B} \mid A)$.

   c. $P(A) = P(B)$.

   d. $P(A) = P(\overline{A})$.

   e. $P(A \cap B) = P(A)P(B)$.

   f. $P(A \cup B) = P(A) + P(B)$.

   g. $P(A \cap B) = P(\overline{A} \cap B)$.

5.1.19. Two containers, labeled 'Right' and 'Left,' are filled with red and blue marbles in the following compositions:

	Left	Right
Red	60	6
Blue	40	4

You choose one of the containers, shake it vigorously, close your eyes, and reach in and grab *two* marbles simultaneously. Suppose two different versions will be played:

(1) You win an attractive prize if *either* marble is *red*.

(2) You win an attractive prize if *either* marble is *blue*.

Which container, Right or Left, gives you the best chances of winning the prize[4] in (1)? In (2)?

    a. Right is best for (1) and Left for (2).

    b. Left is best for (1) and Right for (2).

    c. Left and Right give equal chances of winning in (1) and in (2).

    d. Right is best for (1) and for (2).

    e. Left is best for (1) and for (2).

---

[4]Inspired by a research problem of Cliff Konold.

Chapter 6

Normal Distribution, Sampling Distributions, and Inference

Choose the most correct answer (one and only one) for each problem.

6.1.1. In a normal distribution, the standard score of the tenth percentile (the value that separates the bottom 10 percent from the top 90 percent) is:

    a. -2.33

    b. -0.90

    c. 0.10

    d. -0.10

    e. 0.54

    f. -0.82

    g. -1.28

6.1.2. In a normal distribution of scores with mean 50 and standard deviation 10, which of the following scores is above one-third of the scores and below the other two-thirds?

    a. 33.3

    b. 46.7

    c. −0.43

    d. 43.3

    e. 16.7

    f. 45.7

    g. 62.9

6.1.3. When *increasing* the number of observations in a random sample from the population being studied, we guarantee that:

    a. The distribution of observations in the sample will be approximately normal.

    b. The variance of the observations in the sample will decrease.

    c. The sample's mean will be an unbiased estimator of the population's mean.

    d. The sample's mean will be a random observation from an approximately-normal distribution of means.

    e. The (conditional) probability of Type I error in a statistical test at level of significance 0.05 will decrease.

6.1.4. A certain town is served by two hospitals. In the larger hospital about 45 babies are born each day, and in the smaller hospital about 15 babies are born each day.

(1) As you know, about 50 percent of all babies are boys. However, the exact percentage varies from day to day. Sometimes it may be higher than 50 percent, sometimes lower.

For a period of one year, each hospital records the days when more than 60 percent of the babies born are boys. Which hospital would you expect to record more such days?[1]

   a. The smaller hospital.

   b. The larger hospital.

   c. Both hospitals about the same.

   d. Impossible to tell.

(2) The length of each newborn baby is measured in both hospitals.

For a period of 1 year, each hospital records the days when the *longest* baby born that day exceeds 56 cm. Which hospital would you expect to record more such days?

   a. The smaller hospital.

   b. The larger hospital.

   c. Both hospitals about the same.

   d. Impossible to tell.

Compare parts (1) and (2) of this problem with each other. Make sure you understand the basis for their similarity/difference.

---

[1]Based on Tversky and Kahneman (1974, p. 1125).

6.1.5. We are interested in estimating the *range* of a large and finite population by the range of a random sample (of size 20) from this population. This estimator of the range of the population is which of the following:

    a. An unbiased estimator.

    b. One cannot know whether it is a biased or an unbiased estimator. It depends on the population's distribution.

    c. An upward-biased estimator.

    d. A downward-biased estimator.

6.1.6. In a random sample, when is the sum of squared deviations from the mean, $\sum_{i=1}^{n}(x_i - \bar{x})^2$, divided by $n-1$?

    a. When n is small.

    b. When you are interested in the variance of the sample at hand (comprising n observations).

    c. When the population is not distributed normally.

    d. When you cannot be sure whether the observations are independent of each other.

    e. When you want to get an unbiased estimate of the variance of the population from which that sample has been drawn.

    f. When you wish to compute the variance of the sampling distribution (i.e., the distribution of all sample means of size n).

6.1.7. Consider the proposition that the mean of the means of all samples (sized n) from a given population is equal to the population mean, that is, $\mu_{\bar{x}} = \mu_x$. Which of the following is correct?

    a. The proposition is not true.

    b. The proposition is only approximately true.

    c. The proposition is true only if the population's distribution is normal.

    d. The proposition is true only if n is large.

    e. The proposition is always (exactly) true.

    f. The proposition is true only if the n observations are uncorrelated (e.g., when the sampling is conducted with replacement).

6.1.8. Consider the proposition that $\sigma_{\bar{x}}^2 = \dfrac{\sigma_x^2}{n}$.

    a. The proposition is always (exactly) true.

    b. The proposition is only approximately true.

    c. The proposition is true only if n is large.

    d. The proposition is true only if the n observations are uncorrelated (e.g., when the sampling is conducted with replacement).

    e. The proposition is true only if the population's distribution is normal.

    f. The proposition is not true.

6.1.9.   To estimate an unknown population mean, μ_x, a random sample was taken, and a 95% confidence interval was computed. Let A and B denote (respectively) the lower and upper ends of the confidence interval (obviously $A < B$).

Which of the following statements is *false?*

    a. If we conducted a two-tailed test at level of significance 0.05, based on the present random sample and using any point in the interval AB as μ_0, then H_0 would *not* be rejected. If we used a point not in AB as μ_0, H_0 *would* be rejected.

    b. The sample's mean, $\bar{x}$, is midway between A and B.

    c. The probability that the interval between A and B contains the population mean, μ_x, is 0.95.

    d. The larger the sample size the smaller the distance $B - A$.

    e. The probability that the population mean, μ_x, is midway between A and B is 0.95.

6.1.10.   Scientists conducted an experiment to test two complementary theories T and $\overline{T}$. Before the tests, the two theories were considered equally probable. The experiment resulted in data, denoted D, for which $P(D \mid T) = 1$ and $P(D \mid \overline{T}) = 0.25$.

What can we conclude from these results?

    a. The probability of T should be 1, because the results obtained were predicted with certainty by T.

    b. The probability of T should become four times as great as that of $\overline{T}$.

    c. These findings should not alter our state of knowledge, because obtaining the data D is possible according to both theories.

    d. The posterior probability of theory $\overline{T}$ should be 0.25, and that of T should be 0.75.

6.1.11. The *level of significance*, α, of a statistical test is which of the following:

   a. The probability of rejecting H_0.

   b. The probability of committing an error in the test's result.

   c. The probability that H_0 is true *and* will be rejected.

   d. The (conditional) probability that H_0 is true, given that it was rejected.

   e. The (conditional) probability that H_0 will be rejected, given that it is true.

6.1.12. To obtain a *statistically-significant result* means:

   a. To reject H_0 when it is true.

   b. To get an outcome in the rejection region.

   c. To arrive at a correct conclusion.

   d. To obtain a scientifically-meaningful result.

   e. To obtain a result that will recur in a replication.

6.1.13. A statistical hypothesis is tested at a level of significance of 0.05. If the probability (under H_0) associated with a randomly-observed difference[2] is as small as or smaller than 0.05 then which of the following is true?

    a. The difference is ignored (i.e., considered a deviation due to chance variation).

    b. H_0 cannot be rejected.

    c. The observed difference is not significant.

    d. H_0 is rejected.

    e. A Type I error results.

    f. A Type II error results.

6.1.14. If we conduct a statistical test of a hypothesis, using a random sample, and the result is *not* significant, what conclusion can we draw?

    a. We did not manage to reject H_0.

    b. H_1 should be rejected.

    c. H_0 should be rejected.

    d. H_0 is true.

    e. H_1 is true.

    f. We did not manage to reject H_1.

---

[2]See Chow (1988, p. 108).

6.1.15. In using a statistic s to conduct a test of a statistical hypothesis, which of the following statements about the probability of a Type II error[3] is true?

    a. It depends solely on the sampling distribution of s under H_0.

    b. It depends solely on the sampling distribution of s under H_1.

    c. It is independent of the extent of overlap between the sampling distribution of s under H_0 and of s under H_1.

    d. It is directly related to the extent of overlap between the sampling distribution of s under H_0 and of s under H_1.

    e. It is inversely related to the extent of overlap between the sampling distribution of s under H_0 and of s under H_1.

6.1.16. An investigator designed a statistical test at level of significance 0.05 and determined accordingly the rejection region (and the nonrejection region) for the null hypothesis. She then conducted the experiment on a random sample from the target population. The sample's statistic was computed and found to be in the *rejection region*.

What is the probability that H_0 was erroneously rejected?

    a. 0.05

    b. 0.10

    c. 0.95

    d. 1.645

    e. 0.025

    f. Impossible to know because of missing data.

(See Problem 3.1.10 in Part I)

---

[3]See Chow (1988, p. 106).

6.1.17. An experimenter used a two-tailed test to test a null hypothesis stating that $\mu = \mu_0$. The population is normally distributed, and its variance is unknown. A random sample of 5 observations was employed, and the sample's result was $t(\overline{x}) = -2.611$. What can we conclude about this result?

    a. The result is significant at the 0.10 level and is not significant at the 0.05 level.

    b. The result is significant at the 0.05 level and is not significant at the 0.025 level.

    c. The result is significant at the 0.025 level and is not significant at the 0.01 level.

    d. The result is not significant because its direction is contrary to that of H_1.

Part III

ANSWERS

Answers for Part I: Problems to Solve

Chapter 1. Descriptive Statistics I

1.1 Measures Characterizing Distributions I

1.1.3.   **a.** (1) Increase.

(2) Impossible to know (either increase or no change).

(3) Impossible to know (either increase or no change).

(4) Decrease.

(5) No change.

b. (1) No change.

(2) Impossible to know.

(3) No change.

(4) No change.

(5) Increase.

1.1.4.   **a.** 8, 12

b. 16, 20, 24

c. (1) 6

(2) 3

d. (1) 10

(2) $3\frac{1}{3}$

1.1.5.   There are many solutions. Here are three examples:

$$5, 7, 7, 8, 10, 11, 12, 20.$$

$$2, 6, 7, 7, 11, 14, 16, 17.$$

$$1, 7, 7, 7, 11, 15, 16, 16.$$

1.1.6.   **a.** The mean will *not change.*

b. The variance will *decrease.*

c. The range will *not change.*

d. The symmetry has been used in answering question **c.**

1.1.7.

a.		Optimal suggestion	Fine
	(1)	6	25
	(2)	10	5
	(3)	8	6
	(4)	7	0
	(5)	7	118

b. 16. You get a 'bonus' of 4 points if you have the correct answer and your score is 20 (i.e., excellent!).

1.1.8.   **a.** The median (see Problem 1.1.20).

b. The arithmetic mean.

c. The arithmetic mean.

d. The mode.

e. The median.

f. The midrange, that is, $\dfrac{x_{min} + x_{max}}{2}$.

g. The payment can be made to equal zero, otherwise it might be very large. There is more than one optimal suggestion (the rest is left to the student).

1.1.9.   **a.** 2.0 km. The median of the distances.

   b. 3.6 km. The arithmetic mean of the distances.

   c. 4.0 km. The midrange of the distances.

1.1.11.   6, 8

1.1.12.   **a.** $\bar{x} = 3$, $\sigma_x^2 = 0$.

   b. $\bar{x} = 1$, $\sigma_x^2 = 1$.

   c. $\bar{x} = 2.5$, $\sigma_x^2 = 1.25$.

1.1.13.   **a.** $\frac{1}{2}$

   b. 1, 1, 1, 1, 3, 3, 3, 3

   c. 0

1.1.14.   **a.** 82.69

   b. $M_{n+1} = \dfrac{n\, M_n + x_{n+1}}{n + 1}.$

   c. $d = \dfrac{x_{n+1} - M_n}{n + 1}.$

   d. The mean increases when $x_{n+1} > M_n$. It stays unchanged when $x_{n+1} = M_n$, and it decreases when $x_{n+1} < M_n$.

1.1.15.   **a.** (1) Increase by \$125.

   (2) No change.

   (3) No change.

   (4) Increase by \$125.

(5) Increase by \$125.

b. (1) Multiplied by 1.15.

(2) Multiplied by $1.15^2 = 1.3225$.

(3) Multiplied by 1.15.

(4) Multiplied by 1.15.

(5) Multiplied by 1.15.

c. (1)-(5) None of the measures would change.

1.1.17.   **a.** The density function, $f(x)$, is monotonically decreasing from A to B.

b.
$$
\begin{array}{ccccc}
(1) & < & (2) & > & (3) & > \\
(4) & = & (5) & < & (6) & = \\
(7) & < & (8) & > & (9) & =
\end{array}
$$

1.1.18.

a. $\dfrac{\sum\limits_{i=0}^{7} i f_i}{\sum\limits_{i=0}^{7} f_i} = \dfrac{91}{30} = 3.0.$

b. $\dfrac{\sum\limits_{i=1}^{7} i^2 f_i}{\sum\limits_{i=1}^{7} i f_i} = \dfrac{391}{91} = 4.3.$

c. *Method A* gives the mean number of children *per family*. It answers a question about the size of the *average family*.

Method B gives the mean number of children *per child*. It answers a question about the family size in which the *average child* lives, where "family size" is understood as the number of children in the family.

d. The mean obtained using *Method A* can never be greater than the mean obtained using *Method B*. Usually, the latter will be greater than the former. The two means will be equal only when all the families in the village have the same number of children.

Method B will usually yield a higher result. First, childless families are not taken into account when the children are questioned. (Note that the three families who have no children appear in the denominator of the calculation in **a**, but not in **b**, where only the 91 village children are addressed.) Second, the bigger the family size the more children there are who respond with that bigger number. If we view the formulas in **a** and **b** as weighted means of the family sizes i, we see that in **a** every i is weighted by f_i, whereas in **b** it is weighted by if_i. Consequently, the bigger family sizes are more heavily weighted in *Method B*, thereby increasing the outcome of the calculation (relative to *Method A*).

1.1.19. **a.** The five yearly increases are optimally represented by the average if it replaces each one of them, and the total growth in sales over five years is still the same. After the first increase of 10 percent the sales increase to $100,000 \times 1.10$, a year later, it grows to $100,000 \times 1.10 \times 1.70$, and so on. The sixth year's amount in dollars is therefore,

$$100,000 \times 1.10 \times 1.70 \times 1.25 \times 1.10 \times 1.15.$$

The required average, denoted A, should thus satisfy

$$100,000A^5 = 100,000 \times 1.10 \times 1.70 \times 1.25 \times 1.10 \times 1.15.$$

Hence, $A^5 = 2.9569$. A is thus equal to the fifth root of 2.9569 or approximately 1.2421.

The computed average, 1.2421, is the *geometric mean* of the five numbers 1.10, 1.70, 1.25, 1.10, 1.15. The arithmetic mean of these five numbers is 1.26. The arithmetic mean, however, is an inappropriate average for this situation because $100,000 \times 1.26^5$ does *not* equal the sixth year's sales figure obtained from applying the five yearly increases.

Note that the geometric mean of these five numbers is smaller than their arithmetic mean. This is no accident (see Problem 4.2.6 in Part II). More on the geometric mean as the multiplicative analogue of the arithmetic mean (which is based on addition) can be found in Falk (1984) and Usiskin (1974).

b. (1) 7.5 kilometers per hour.

This average is the *harmonic mean* of the numbers 5 and 15. If an object travels from point A to point B with velocity v_1, and from B to A with velocity v_2, then the object's average velocity v, over the whole trip, is the harmonic mean of v_1 and v_2, namely,

$$v = \frac{2}{\frac{1}{v_1} + \frac{1}{v_2}} = \frac{1}{\frac{\frac{1}{v_1} + \frac{1}{v_2}}{2}} .$$

(Make sure you can derive this solution.) The harmonic mean is thus the reciprocal of the arithmetic mean of the reciprocals of the averaged numbers. It is never greater than the geometric mean of the same numbers (assuming the numbers are positive). A detailed exposition of the properties of various averages and the order relations between them can be found in Bullen (1990), Burrows and Talbot (1986), Hoehn and Niven (1985), and Maor (1977).

(2) No matter how fast the skier travels down, he cannot attain an average speed that will double his speed going up. To double his speed, he would have to go *twice* the original distance (of the slope) in the same length of time it took him to go up. This leaves no time at all for the downward trip (Gardner, 1982, p. 142).

Formally, the skier's wish can be expressed as follows:

$$\frac{2}{\frac{1}{v_1} + \frac{1}{v_2}} = 2v_1 \quad .$$

This equality implies that $v_2 = v_2 + v_1$, which is impossible, since we already know that $v_1 \neq 0$.

In more general terms, the harmonic mean of two positive numbers, a and b, is always less than $2a$ and less than $2b$. Let $H(a, b)$ denote the harmonic mean of a and b, then

$$H(a, b) = \frac{2}{\frac{1}{a} + \frac{1}{b}} = \frac{2ab}{a + b} \quad .$$

It is easy to see that

$$\frac{2ab}{a + b} < 2a \text{ and } \frac{2ab}{a + b} < 2b \quad .$$

Note that no similar property holds for the arithmetic mean or for the geometric mean of two positive numbers.

1.1.20. This theorem is widely known, yet its proof rarely appears in standard textbooks. It is considered difficult, as is usually the case when the absolute-value function is involved. There do exist, however, some simple proofs (e.g., Arbel, 1985; Buchart & Moser, 1952; Joag-Dev, 1989; Kokan, 1975; Schwertman, Gilks, & Cameron, 1990; Yule & Kendall, 1953, p. 138). We briefly present the proof that appears in most of them.

Proof. Without loss of generality, we may assume that

$$x_1 \leq x_2 \leq x_3 \leq \ldots \leq x_n \quad .$$

We can visualize the x's on the number line:

Suppose we wish to minimize the sum of distances (i.e., absolute deviations) just to x_1 and x_n. It is clear that any point A that lies between x_1 and x_n will do the job. Thus, $|\,x_1 - A\,| + |\,x_n - A\,|$ is minimal when A lies in the interval $[x_1, x_n]$. By the same reasoning, $|\,x_2 - A\,| + |\,x_{n-1} - A\,|$ is minimal when A lies in the interval $[x_2, x_{n-1}]$. Similarly, $|\,x_3 - A\,| + |\,x_{n-2} - A\,|$ is minimal when A lies in $[x_3, x_{n-2}]$, and so on.

The sum of all the absolute deviations, $\sum_{i=1}^{n} |\,x_i - A\,|$, will be minimal when A lies in the *intersection* of all the above intervals. This is so because if A lies out of this intersection, some of the $\dfrac{n+1}{2}$ or $\dfrac{n}{2}$ terms (depending on whether n is, respectively, odd or even), which correspond to the above intervals, increase while none of the other terms decrease. The intersection of all these nested intervals, however, equals the median point if n is odd, and the median interval if n is even. This completes the proof for every n.

Note. A specific case of an extension of this theorem to two dimensions, with a surprisingly simple proof, appears in Honsberger (1978, pp. 1–2).

1.1.21.   To simplify matters, let's go over to a set of standard scores $z_i = (x_i - M)/\sigma$ (for $i = 1, 2, \ldots, n$), which has mean 0 and standard deviation 1. Therefore,

$$\sum_{i=1}^{n} z_i = 0 \qquad \text{and} \qquad \sum_{i=1}^{n} z_i^2 = n \quad . \tag{1}$$

The median of the z's, denoted z^*, is $z^* = (Me - M)/\sigma$. Our objective is to show that

$$|z^*| \le 1 \quad . \tag{2}$$

We shall prove (2) for the case when n is even and let the reader complete the proof for n odd (the method requires only minor modification).

Suppose that $|z^*| > 1$,

If $z^* > 1$, then $z_i > 1$ for each $i > n/2$, and therefore

$$\sum_{i=(n/2)+1}^{n} z_i > \frac{n}{2} \quad \text{and} \quad \sum_{i=(n/2)+1}^{n} z_i^2 > \frac{n}{2} \quad . \quad (3)$$

Equalities (1) and (3) imply that

$$\sum_{i=1}^{n/2} z_i < -\frac{n}{2} \quad \text{and} \quad \sum_{i=1}^{n/2} z_i^2 < \frac{n}{2} \quad . \quad (4)$$

Because of (4), the variance σ_1^2 of $z_1, z_2, \ldots, z_{n/2}$,

$$\sigma_1^2 = \frac{\sum_{i=1}^{n/2} z_i^2}{n/2} - \left[\frac{\sum_{i=1}^{n/2} z_i}{n/2} \right]^2,$$

is negative. This, of course, is impossible.

If $z^* < -1$, similar reasoning leads to the impossible conclusion that $\sigma_2^2 < 0$, where σ_2^2 is the variance of the values $z_{(n/2)+1}, z_{(n/2)+2}, \ldots, z_n$. Thus (2) is true. The equality $|z^*| \le 1$ is equivalent, however, to $|M - Me| \le \sigma$.

Since the absolute gap between the mean and the median is bounded by the standard deviation, the measure of skewness (asymmetry) based on the difference between the mean and the median is computed relative to the standard deviation. Consequently, that index of skewness satisfies the inequality:
$-1 \le \dfrac{M - Me}{\sigma} \le 1$.

1.1.22. We present a brief proof by Sher (1979) that we consider elegant (see also Page & Murty, 1982, 1983).

Given any distribution $x_1 \le x_2 \le \ldots \le x_n$, we note that $x_n = x_1 + R$. Let $y_i = x_i - (x_1 + R/2)$ so that

$$-R/2 = y_1 \le y_2 \le \ldots \le y_n = R/2.$$

The x's and y's have the same standard deviation since x is transformed into y by subtracting a constant. Let σ be that standard deviation ($\sigma = \sigma_x = \sigma_y$). We have

$$\sigma = \sqrt{(\sum_{i=1}^{n} y_i^2)/n - M_y^2} \leq \sqrt{(\sum_{i=1}^{n} y_i^2)/n} \leq \sqrt{n(R/2)^2/n} = R/2.$$

This proves that $\sigma \leq R/2$.

It is easy to see that, for a given positive R, the maximal possible standard deviation, $\sigma = R/2$, is obtained when the distribution comprises only two different values and is symmetric. That is, the observations are equally divided between x_{min} and x_{max}. Obviously, $x_{max} - x_{min} = R$.

1.2 Correlation and Linear Regression

1.2.2.   **a.** $r = 1$.

b. $r_s = 1$.

c. The linear correlation coefficient, r, will decrease, but remain positive (and quite high). The rank-order correlation coefficient, r_s, will remain equal to 1.

1.2.3.   **a.** (1) The linear correlation coefficient will not change.

(2) The rank-order correlation coefficient will not change.

b. (1) The linear correlation coefficient will change in value, but in most cases its sign will not change.

(2) The rank-order correlation coefficient will not change.

1.2.4.   **a.** In case (2).

b. (1) 49%

(2) 81%

c. (1) 49%

(2) 81%

1.2.5. Answer (4) is the correct one.

Since the correlation coefficient is positive but less than 1, linear predictions remain on the same side of the mean but 'regress' to it (in terms of standard-deviation units). Starting from a theoretical score of 83, the predicted practical score would still be above the mean (of the practical scores), but closer to it in standard-deviation units than was 83 to 75. The same is true for linearly predicting a theoretical score starting from the predicted practical score. Thus, the original score regresses twice to the mean, yet stays above it. It will therefore be greater than 75 and less than 83.

Note that we are given sufficient data to exactly compute the target score. It will be $75 + 0.65^2(83 - 75) = 78.38$.

1.2.6. In the answers below, Y means that the pair of functions *could* be regression lines, and N means that they could *not*.

  a. N (The product of the regression coefficients must always satisfy $0 \leq b_1 b_2 \leq 1$.)

  b. Y

  c. N (The two regression coefficients must always have the same sign.)

  d. N (These are nonlinear functions.)

  e. Y

  f. N (The first regression line indicates a zero correlation, whereas the second implies a positive correlation.)

  g. N (The product of the regression coefficients and their signs imply that the correlation is 1, but the two equations do not define the same line.)

1.2.7. The solution depends on which of the two regression lines is used to make the required prediction. Since we are given a test score and asked to predict age, we should use the regression line of ages on test scores.[4] The predictor is the score 81, and the information about the child's chronological age (9 years) is superfluous. The answer is $11.6 + \frac{2.72}{9.23}0.61(81 - 79.5) = \mathbf{11.9}$.

Chapter 2. Probability I

2.1 Events, Operations, and Event Probabilities

2.1.1.   In the list below, T denotes the assertion is true, and F denotes it is false.

a.	F	**d.**	F	**g.**	T	**j.**	T
b.	T	**e.**	T	**h.**	F	**k.**	T
c.	F	**f.**	T	**i.**	T	**l.**	T

2.1.3.   The probability rankings of the four categories described in the problem violate the conjunction rule, which is a simple and most basic law of probability (Tversky & Kahneman, 1983). The law states that the probability of a conjunction, $P(A \cap B)$, cannot exceed the probabilities of its constituents, $P(A)$ and $P(B)$, because the conjunction event (or set) is always included in its constituents. No matter what the definitions of A and B are, it is always true that $P(A \cap B) \leq P(A)$ and $P(A \cap B) \leq P(B)$. That is so because $A \cap B \subseteq A$ and $A \cap B \subseteq B$.

---

[4]The other regression line, namely, that of scores on ages, might *seem* more appropriate, since age is usually employed as an independent variable with performance as dependent variable. Hence, 'predicting age' appears an unnatural task. Despite that psychological difficulty, it would be erroneous to insert the score 81 in the regression of scores on ages and solve for the age predicting that score. Predicting age via the regression of ages on scores is correct in the sense of using the line for which the sum of squared errors in prediction of ages is minimal. (See Oakley & Baker, 1977 and a response to that paper by Falk, 1979b.)

In the employment problem, category **d** (unemployed) presents one constituent of a conjunction event defined in category **b** (married *and* unemployed). It is therefore *incorrect* to rank order the conjunction (**b**) as more probable than the constituent (**d**), regardless of the employment statistics for academc women in the social sciences.

People's judgments under uncertainty, however, are often mediated by intuitive heuristics that are not bound by the conjunction rule. A conjunction may appear as more plausible than one of its constituents because it seems more coherent or representative of some image of the target event. "Married and unemployed" may make more sense (than just "unemployed") because being a married woman (having a family) can *causally* account for being unemployed.

2.1.4. **a.** $P(T) = P(H) = 1/2$. The procedure is fair.

b. $P(T) = 1/4$ and $P(H) = 3/4$. The procedure is biased in favor of Harriet.

c. $P(T) = 5/9$ and $P(H) = 4/9$. The procedure is biased in favor of Tom.

d. $P(T) = P(H) = 18/36$. The procedure is fair.

2.1.5. **a.** $a + b$

b. 0

c. $1 - a$

d. b

e. 1

f. $1 - (a + b)$

g. a

2.1.6. **a.** b **d.** $b-a$ **g.** 1
            b. $1-a$ **e.** 0 **h.** $1-b$
            c. a **f.** $1-b$ **i.** $b-a$

2.1.7. **a.** $P(A \cup B) = 0.45$.
            b. $P(A \cup B) = 0.60$.
            c. $P(A \cup B) = 0.40$.
            d. $P(A \cup B) = 0.55$.
            e. $P(A \cup B) = 0.60$.
            f. $P(A \cup B) = 0.40$.

2.1.8. **a.** $H = A \triangledown S = (A \cup S) - (A \cap S)$, where $\triangledown$ denotes the 'symmetric difference.'

Let E and F be two sets (events) in a sample space Ω. Their *symmetric difference*, $E \triangledown F$, is defined as the set of all elements that belong to *one and only one* of the sets E or F (see the Venn diagram). Note that

$$E \triangledown F = (E \cup F) - (E \cap F) = (E \cap \overline{F}) \cup (\overline{E} \cap F).$$

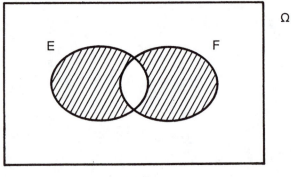

E ∇ F

b. (1) 0.54   (2) 0.82

2.2 Problems Involving Combinatorics

2.2.1.   **a.** $\frac{1}{3^{13}} = 627 \times 10^{-9}$.

   b. $\frac{1 + 2 \times 13}{3^{13}} = \frac{27}{3^{13}} = 169 \times 10^{-7}$.

2.2.2. $\frac{9}{24} = 0.375$ (See also answer to Problem 2.5.13b)

2.2.3. $\dfrac{1}{\dbinom{8}{4}} = \frac{1}{70} = 0.014$.

2.2.4.   **a.** $\frac{1}{(8)_3} = 0.0030$.

   b. $\dfrac{1}{\dbinom{8}{3}} = 0.018$.

2.2.5.   **a.** $\frac{1}{6!} = \frac{1}{720} = 0.00139$.

   b. $\frac{1}{(6)_3} = \frac{1}{120} = 0.00833$.

   c. $\dfrac{1}{\dbinom{6}{3}} = \frac{1}{20} = 0.0500$.

Note. It is easy to see why the number of permutations of six distinct figures is 6!. It is, however, less obvious why the number of distinct permutations reduces to $(6)_3$ when three of the six digits are identical, and why it further reduces to $\binom{6}{3}$ when the other three figures become indistinguishable as well. On the next page is a "proof without words" of similar formulas, starting with four distinguishable symbols.

Permutations	of four	objects
$4! = 24$	$(4)_2 = \frac{4!}{2!} = 12$	$\binom{4}{2} = \frac{4!}{2!2!} = 6$
♠ ♣ ♡ ♢	♣ ♣ ♡ ♢	♣ ♣ ♡ ♡
♣ ♠ ♡ ♢		
♠ ♣ ♢ ♡	♣ ♣ ♢ ♡	
♣ ♠ ♢ ♡		
♠ ♡ ♣ ♢	♣ ♡ ♣ ♢	♣ ♡ ♣ ♡
♣ ♡ ♠ ♢		
♠ ♢ ♣ ♡	♣ ♢ ♣ ♡	
♣ ♢ ♠ ♡		
♡ ♠ ♣ ♢	♡ ♣ ♣ ♢	♡ ♣ ♣ ♡
♡ ♣ ♠ ♢		
♢ ♠ ♣ ♡	♢ ♣ ♣ ♡	
♢ ♣ ♠ ♡		
♡ ♠ ♢ ♣	♡ ♣ ♢ ♣	♡ ♣ ♡ ♣
♡ ♣ ♢ ♠		
♢ ♠ ♡ ♣	♢ ♣ ♡ ♣	
♢ ♣ ♡ ♠		
♡ ♢ ♠ ♣	♡ ♢ ♣ ♣	♡ ♡ ♣ ♣
♡ ♢ ♣ ♠		
♢ ♡ ♠ ♣	♢ ♡ ♣ ♣	
♢ ♡ ♣ ♠		
♠ ♡ ♢ ♣	♣ ♡ ♢ ♣	♣ ♡ ♡ ♣
♣ ♡ ♢ ♠		
♠ ♢ ♡ ♣	♣ ♢ ♡ ♣	
♣ ♢ ♡ ♠		

2.2.6. Suppose the dice are colored white, grey, and black (Freedman et al., 1978, p. 222). We consider all possible different outcomes of rolling these three dice. There are $6^3 = 216$ ways that three distinguishable dice can fall. Each of these outcomes is equally likely if the dice are fair and are rolled properly. We must count how many of the possible outcomes sum to 9 and how many sum to 10, in order to compute the two probabilities.

The trouble with the Italian gamblers' argument is that they only counted triplets, and they failed to consider the different ways to obtain a given triplet. For instance, the triplet 3 3 3, (with total 9) can be obtained in only one way, whereas the triplet 3 3 4, (with total 10) can be obtained in three different ways, since 4 dots can occur on either the white, the black, or the grey die.

The exact count of all the ways of getting 9 or 10 dots with three dice is given below:

Triplets for 9	Number of ways to roll	Triplets for 10	Number of ways to roll
1 2 6	6	1 4 5	6
1 3 5	6	1 3 6	6
1 4 4	3	2 2 6	3
2 3 4	6	2 3 5	6
2 2 5	3	2 4 4	3
3 3 3	1	3 3 4	3
Total	25	Total	27

There are 25 ways to get a total of 9 and 27 ways to get a total of 10. The probability of rolling 9 is thus $25/216 = 0.116$, and that of rolling 10 is $27/216 = 0.125$.

2.2.7.   **a.** $\dfrac{1}{\binom{20}{5}} = 0.64\bar{5} \times 10^{-4}$.

b. $\dfrac{\dbinom{15}{5}}{\dbinom{20}{5}} = 0.194.$

2.2.8.   **a.** $1 - \dfrac{(365)_{10}}{365^{10}} = 0.117.$

b. $1 - \dfrac{364^{10}}{365^{10}} = 0.0271.$

The answer to **a** is usually surprising. Most people expect it to be much lower. Note that if there were 23 people at the gathering, the answer to question **a** would be 0.507, and the answer to question **b** would be 0.0612.

2.2.9.   **a.** $\dfrac{5!}{5^5} = 0.038.$

b. $\dfrac{3!}{5^3} = 0.048.$

2.2.10. a. $\dfrac{1}{(10)_4} = \dfrac{1}{10 \cdot 9 \cdot 8 \cdot 7} = 0.000198.$

b. $\dfrac{1}{\dbinom{11}{3,2,2,2}} = \dfrac{3!\,2!\,2!\,2!}{11!} = 1.202 \times 10^{-6}.$

2.2.11.   **a.** (1) $\dfrac{1}{31^3} = \dfrac{1}{29,791} = 0.336 \times 10^{-4}.$

(2) $\dfrac{1}{(31)_3} = \dfrac{1}{31 \times 30 \times 29} = \dfrac{1}{26,970} = 0.371 \times 10^{-4}.$

b. (1) $\dfrac{1}{\dbinom{33}{3}} = \dfrac{1}{5,456} = 0.183 \times 10^{-3}.$

(2) $\dfrac{1}{\dbinom{31}{3}} = \dfrac{1}{4,495} = 0.222 \times 10^{-3}.$

Compare with the answers to Problem 2.2.12, and see also the discussion there.

2.2.12.   **a.** $\dfrac{4^3}{5^3} = 0.512.$

b. $\dfrac{(4)_3}{(5)_3} = 0.400.$

c. $\dfrac{\dbinom{6}{3}}{\dbinom{7}{3}} = 0.571.$

d. $\dfrac{\dbinom{4}{3}}{\dbinom{5}{3}} = 0.400.$

Discussion. Our main lesson from consideration of the four sections of the problem, and the different answers that they yield, is that a statement like that of Dr. Zed telling his guests that he had asked his wife to distribute the coins "at random" in the boxes, is not sufficiently precise. A problem in probability can be unequivocally interpreted only when the *statistical experiment*, which the solver purports to analyze, is strictly defined (see Problems 2.4.11, 2.4.14, answers and discussions). Describing a process as being conducted "at random" might (sometimes) be meaningless, unless one fully specifies how the process is carried out. (For example, four methods of randomly picking three of 31 ice cream flavors are described in Problem 2.2.11.)

When distributing objects in containers, we need to clarify whether or not different objects are distinguishable, so that we know whether swapping two objects in different containers yields a new arrangement. Likewise, we need to spell out other terms of the process, including whether only one object is allowed per container or whether we can put many objects in a container. Only after we have made these rules explicit, can we unequivocally interpret the information that the objects have been distributed "at random" to mean that all possible arrangements conforming to the specified rules are equally likely.

By the same token, in problems of sampling, we must specify whether the order in which items are sampled matters or whether we only care about which set of items is sampled. Likewise, we need to clarify whether the samping is conducted with or without replacement.

By now, the reader has probably sensed the analogy between the variables defining a sampling process and those characterizing a procedure of distributing objects in containers (as exhibited in the analogy between the respective four kinds of questions in Problem 2.2.11 and in the present problem). Just as drawing samples of size r from a population of size n can be divided into four types of procedures, distributing r objects among n containers can be classified into four (respectively) *isomorphic* procedures.

Sampling with order parallels distributing distinguishable objects into containers, whereas unordered sampling corresponds to distributing indistinguishable objects. Likewise, sampling with replacement corresponds to allowing more than one object per container, whereas sampling without replacement corresponds to allowing only one object per container. No wonder there is *one* combinatorial formula for each *pair* of isomorphic problems (one from each context). See the two tables on the next page:[5]

---

[5]See also Hamdan (1978), Kreith and Kysh (1988), Stanley (1986, p. 31), and Troccolo (1977).

Number of samples of size r that can be drawn from of a set of n objects, under various sampling conditions.

	Ordered	Unordered
With replacement	n^r	$\binom{n+r-1}{r}$
Without replacement $(r \leq n)$	$(n)_r$	$\binom{n}{r}$

Number of ways to distribute r objects among n containers, under various distribution conditions.

	Distinguishable objects	Indistinguishable objects
No limitations	n^r	$\binom{n+r-1}{r}$
No more than one object per container $(r \leq n)$	$(n)_r$	$\binom{n}{r}$

Note that

$$(m)_k = m(m-1)(m-2)\ldots(m-k+1)\,,$$

and $\displaystyle \binom{m}{k} = \frac{m(m-1)(m-2)\ldots(m-k+1)}{1 \times 2 \times 3 \times \ldots \times k}\,.$

The derivation of these formulas is straightforward and can be found in almost any textbook on probability or combinatorial analysis. The hardest of the four (which is consequently less often presented in textbooks) is the case of the number of *un-ordered* samples of size r that can be drawn *with replacement* from a set of n objects. We will now prove the formula for the number of ways to distribute r *indistinguishable* objects among n containers with *no limitations*. It suffices to prove the formula for only one of the two problems because they are isomorphic.[6]

Suppose three dimes are placed in Dr. Zed's five boxes, as depicted:

Without loss of generality, we may move the boxes close to each other and draw just one thick line for two adjacent sides:

When n boxes are lined up this way, there are always $n-1$ thick vertical lines which distinguish the separate boxes. A given distribution of r objects in n boxes can be uniquely described by a sequence of r circles (representing the objects) and $n-1$ vertical lines (representing divisions between boxes). Thus, for example, the above distribution is described by $|\,0\,|\,|\,0\,0\,|$. This sequence starts and ends with a $|$ because the first and last boxes are empty. The two consecutive vertical lines in the third and fourth places represent the empty third box, and the two consecutive circles in the fifth and sixth places represent the two dimes in the fourth box.

---

[6]A discussion of that case can be found in Dessart (1971), Feller (1957, pp. 36-37), Golomb (1968), Gupta (1955), Kreith and Kysh (1988), and Troccolo (1977).

Conversely, every sequence of $n - 1$ vertical lines and r circles defines one distribution of r dimes in n boxes. For example, 0 | | | | 0 0 defines the following distribution of three dimes in five boxes:

Because of the one-to-one correspondence between distributions of r dimes in n boxes and sequences of $n - 1$ vertical lines and r circles, the required number of distributions equals the number of distinct sequences of the above type. But we know that the number of permutations of $n - 1$ indistinguishable symbols of one kind and r indistinguishable symbols of another kind is $\begin{pmatrix} n + r - 1 \\ r \end{pmatrix}$ (see Problem 2.2.5.c, the answer to this problem, and the illustration on page 148).

In solving probability problems, it is of utmost importance to carefully consider the experimental procedure defining the problem before rushing to apply combinatorial formulas. If, for example, the sampling with replacement of r objects from a set of n objects had been conducted in an ordered manner, so that in sampling the ith object all n outcomes were equiprobable, for each i, it would *not* be justified to assume that all $\begin{pmatrix} n + r - 1 \\ r \end{pmatrix}$ unordered samples have the same probability. A sample containing object a three times would be less probable than a sample comprising objects a, b, and c. Not all triplets will be equiprobable under that experimental procedure. (An error of carelessly assigning equal probabilities to different kinds of triplets had been committed by seventeenth century gamblers, as described in Problem 2.2.6.)

On the other hand, in the case of Mrs. Zed distributing coins in boxes, we are told from the beginning that she acts in a way guaranteeing "that any possible arrangement has an equal chance of being the one selected." Therefore, in case **c** (as described by Professor Gamma) the probability of any possible specific distribution is $1/\binom{5+3-1}{3}$ or $1/\binom{7}{3}$, that is, $1/35$.

In games of chance, and when generating randomness for experimental purposes and other human affairs, one usually operates so that all n^r ordered samples (with replacement) are equiprobable. Thus, when disregarding order, it would be wrong to assume that all $\binom{n+r-1}{r}$ samples have the same probability. In the physical world, however, the situation is reversed: the former circumstances are virtually nonexistent, whereas the latter describe some phenomena in atomic physics.

In statistical mechanics (see Feller, 1957, p. 39; Meshalkin, 1963/1973, pp. 11–13; and Troccolo, 1977) one usually subdivides the phase space into a large number n of small regions or cells so that each of r indistinguishable particles falls into one of the cells. The state of the entire system is uniquely described by the distribution of the r particles into the n cells. Offhand it would seem that all n^r arrangements would have equal probabilities. If this is true, the physicist speaks of *Maxwell-Boltzmann statistics* (Professor Alpha's postulation). Numerous attempts have been made to prove that physical particles behave in accordance with Maxwell-Boltzmann statistics, but modern theory has shown beyond doubt that these statistics do not apply to any known particles; in no case are all n^r arrangements approximately equally probable.

If all possible arrangements of r indistinguishable particles in n cells — when no limitations are imposed on the number of particles per cell — are equally probable, the physicist speaks of *Bose-Einstein statistics* (Professor Gamma's postulation). In this case the probability of every distinct arrangement is $\binom{n+r-1}{r}^{-1}$. It has been found empirically that photons, atomic nuclei, and atoms containing an even number of elementary particles (such as hydrogen atoms) are subject to Bose-Einstein statistics.

The model in which the r particles are indistinguishable, but no more than one particle may occupy a given cell, yields $\binom{n}{r}$ distinct arrangements; if all of them have equal probabilities, the physicist speaks of *Fermi-Dirac statistics* (Professor Delta's postulation). It has been shown in statistical mechanics that this assumption applies to electrons, neutrons, and protons.

These examples highlight the impossibility of selecting or justifying probability models on a priori grounds. In fact, no one could use pure reasoning to predict that photons and protons would not obey the same probability laws. One must resort to experimentation in order to find which model best represents the behavior of the particles in question. Having found these results experimentally, we may seek a reason for them on theoretical grounds. Thus, for example, it was shown that elementary particles do not possess an 'individuality.' It is not possible to think of them as labeled or distinguishable in any way. This situation immediately rules out Professors Alpha's and Beta's suggestions as descriptions of these particles' behavior.

In conclusion, the lesson for the student of probability is that figuring out the number of outcomes is not enough when you wish to find event probabilities. The justification of using combinatorial results for probability computations usually depends on the uniformity of our (finite) discrete sample-space. Uniformity, however, is not automatically guaranteed. It has either to be assumed (when it makes sense . . .) or to be empirically verified.

2.2.13. **a.** $\dfrac{2}{\dbinom{10}{5}} = 0.00794.$

b. Same as **a.**

c. $\dfrac{5!\,2^5}{10!} = \dfrac{10 \times 8 \times 6 \times 4 \times 2}{10!} = \dfrac{1}{9} \cdot \dfrac{1}{7} \cdot \dfrac{1}{5} \cdot \dfrac{1}{3} \cdot 1 = 0.00106.$

Note that each of the three expressions in part **c** represents a different way of reasoning which leads to the same answer.

2.2.14. For a palindrome to occur, the first digit should equal the last — probability $1/10$, and the second digit should equal the penultimate (second-to-last) digit — probability $1/10$. These two events are independent (bringing us to the topic of the next chapter). Therefore, the probability of the required event is $(1/10)(1/10) = 1/100$.

The above reasoning can also be presented in combinatorial terms. It is somewhat more complicated, however.

2.3 Conditional Probabilities and Dependence/Independence between Events

2.3.1. $P(T) = 8/118 = 0.068;\ \ P(T \mid \mathit{1st}) = 3/29 = 0.103.$

The inequality between these probabilities reflects an increased tendency of English words to start with t, relative to the general rate of appearance of that letter (note also this very sentence . . .).

2.3.2. $P(H) = 2/49 = 0.041.$

$P(H \mid T) = 2/5 = 0.400.$

The inequality $P(H \mid T) > P(H)$ reflects an increased tendency of the letter h to come after t, relative to h's general rate of appearance.

2.3.3.

No. of child	Probability of being selected
1	$\frac{1}{6}$
2	$\frac{5}{6} \times \frac{1}{5} = \frac{1}{6}$
3	$\frac{5}{6} \times \frac{4}{5} \times \frac{1}{4} = \frac{1}{6}$
4	$\frac{5}{6} \times \frac{4}{5} \times \frac{3}{4} \times \frac{1}{3} = \frac{1}{6}$
5	$\frac{5}{6} \times \frac{4}{5} \times \frac{3}{4} \times \frac{2}{3} \times \frac{1}{2} = \frac{1}{6}$
6	$\frac{5}{6} \times \frac{4}{5} \times \frac{3}{4} \times \frac{2}{3} \times \frac{1}{2} \times 1 = \frac{1}{6}$

The procedure is certainly *fair*, because before starting each participant has an equal chance of being selected. The fairness of selecting one winner via such a *without replacement* draw is apparently not obvious. The poet J. W. Goethe tells in his autobiographical account in *Poetry & Truth: From My Own Life* about his grandfather's nomination as *Schultheiss* which was decided by a draw. The lottery for selecting one of three candidates consisted of their representatives drawing balls in turn (without replacement) from a bag with two silver and one golden ball. The winner was the one drawing the golden ball. Goethe notes that

> What rendered the circumstance particularly re-markable was, that although his representative was the third and last to draw at the balloting, the two silver balls were drawn first, leaving the golden ball at the bottom of the bag for him.

It appears that Goethe believed that being the last is least favorable.

2.3.4.   **a.** $l_{70} = l_{65}(1 - 0.160) = 0.627.$

        b. $1 - (l_{55}/l_{50}) = 0.03\dot{5}.$

        c. The missing datum is the value of l_{25}. The (conditional) probability of interest for that student is l_{50}/l_{25}.

        d. 0.947

2.3.5.   **a.** $1 - \left(\frac{5}{6}\right)^4 = 0.518.$

        b. $1 - \left(\frac{35}{36}\right)^{24} = 0.491.$

2.3.6. $p = 1 - (1 - p_1)(1 - p_2)\ldots(1 - p_n).$

2.3.7. $0.42^2 + 0.33^2 + 0.18^2 + 0.07^2 = 0.323.$

2.3.8.   **a.** 0.41

        b. 85%

2.3.9.   He should put one white ball in one jar, and the rest of the balls (199) in the other jar.

2.3.10.

Faculty	Men - m		
	Number of appli- cants	Admission data (A)	
		Number	Probability
F_1	775	349	0.45
F_2	225	11	0.05
Total	1000	360	0.360

Faculty	Number of appli- cants	Women - w	
		Admission data (A)	
		Number	Probability
F_1	100	90	0.90
F_2	900	90	0.10
Total	1000	180	0.180

2.3.11. The princess should wait in room B, since

$$P(A) = \frac{1}{3} \times \frac{1}{2} + \frac{1}{3} \times \frac{1}{2} \times \frac{1}{2} + \frac{1}{3} \times \frac{1}{3} + \frac{1}{3} \times \frac{1}{3} = \frac{17}{36},$$

and

$$P(B) = \frac{1}{3} \times \frac{1}{2} \times \frac{1}{2} + \frac{1}{3} + \frac{1}{3} \times \frac{1}{3} = \frac{19}{36},$$

where A and B denote the respective events that Reynaldo will enter room A or B.

2.3.12. This problem provoked much dispute and confusion in its time. Contradictory answers created the legend of a paradox. Solutions involving proportional reasoning — such as dividing the stakes in the ratio of rounds won, 5 : 3, or in the ratio $(6 - 3) : (6 - 5)$ — were suggested (as was the case for the first de Méré problem — Problem 2.3.5). Pascal and Fermat considered it a problem of *probabilities*. They ruled that a fair division should match the ratio of the probability of player A winning to that of player B winning, if they were to continue playing.

These probabilities can easily be derived using the 'tree diagram' on the next page (the current count of rounds won is written in the circles; the outcomes are written in the squares):

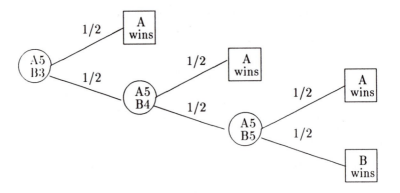

Let A and B denote respectively the events that player A or player B wins. It follows that

$$P(A) = \frac{1}{2} + \frac{1}{2} \cdot \frac{1}{2} + \frac{1}{2} \cdot \frac{1}{2} \cdot \frac{1}{2} = \frac{7}{8},$$

and

$$P(B) = \frac{1}{2} \cdot \frac{1}{2} \cdot \frac{1}{2} = \frac{1}{8}.$$

The stakes should be divided in the ratio 7 : 1 in favor of A.

2.3.13. 0.74

2.3.14. No difference in the number of sisters should be expected between men and women, and in both cases the expected number of sisters is equal to that of brothers.

Discussion. Despite accepting the assumptions of equal probabilities for male and female births and of independence among births, most people expect men to have more sisters than women have and to have more sisters than brothers. It seems that because, on the average, families have an equal number of sons and daughters, the set of siblings of men, who are themselves not counted, should comprise an excess of women (sisters). This apparently compelling reasoning is nevertheless wrong! Men are expected to have the same number of sisters as women have and to have the same number of sisters as brothers.

The correct analysis of this problem requires a full realization of the significance of *statistical independence*. Asking a random child from a family of n children how many brothers and how many sisters he or she has, is equivalent (mathematically) to picking a random family with $n-1$ children and asking how many sons and how many daughters it has. In the latter case it is obvious, however, that one should expect the same number of sons and daughters. Hence, irrespective of whether the questioned persons are males or females, they would have, on the average, the same numbers of brothers and sisters.

Even though the above reasoning is right, some doubt may still linger in readers' minds. The feeling may persist that men must have more sisters since the men are not counted in the statistics of their families. That doubt may be enhanced by the apparent similarity to sampling *without* replacement from an urn with half Ms and half Fs. It is important, therefore, to dwell further on this problem until the reader clearly sees that asking a man about the sexes of his siblings is — statistically speaking — identical to asking a woman the same question. This is the *essence* of the concept of *statistical independence*: the probabilities of male or female childbirths in the respondent's family are *not affected* by knowledge of the respondent's sex.

Suppose we select one specific random outcome in a sequence of (independent) coin tosses and randomly delineate a subsequence of some finite size around it. If asked how many heads and tails are expected among the *other* outcomes in that subsequence ('family'), the answer would certainly be that equal numbers of the two outcomes are expected. That answer would be valid irrespective of whether the selected outcome was heads or tails. Unlike coin tosses, however, childbirths are naturally grouped in families, and that may be the cause of some of the confusion. But there is no more dependence between the sexes of the newborns in the same family than between the outcomes in different coin tosses.

Some readers may nevertheless experience some conflict between the long-run expectation of equal proportions of males and females and the exclusion of the respondents from consideration in a finite population. The following is a representative argument that was given by a male student in one of our classes:

> Suppose there were 1,000 men in our class. Consider all of us together with our brothers and sisters. That large group should comprise approximately equal proportions of males and females. Now, exclude us, 1,000 men, and there would remain more sisters than brothers.

This reasoning fails to take into account the fact that all-daughter families are not represented among those 1,000 families, and, moreover, that families with many sons are overrepresented in that large group. The group will therefore contain an excess of males. Only after removing the original 1,000 men should the expected number of males (brothers) equal that of females (sisters). Incidentally, the above argument provides an interesting example of the bias introduced by sampling families via their sons. (See *Method B* in Problem 1.1.18 for a somewhat similar situation.)

It is noteworthy that Galton (1869), in *Hereditary Genius* (p. 122) erred in concluding that since judges were all men and came from families of average size five, each must have (on average) $2\frac{1}{2}$ sisters and $1\frac{1}{2}$ brothers. Later, Galton (1904) realized his error.

2.3.15.   **a.** The proof is presented in the figure below (remember that $P(A) \neq 0$ and $P(B) \neq 0$):

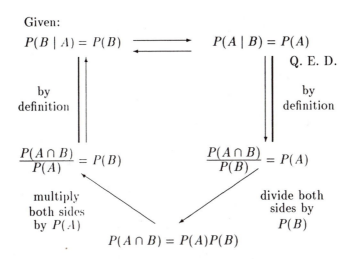

Given:

$P(B \mid A) = P(B)$ $\longrightarrow$ $P(A \mid B) = P(A)$

Q. E. D.

by definition   by definition

$\dfrac{P(A \cap B)}{P(A)} = P(B)$ $\dfrac{P(A \cap B)}{P(B)} = P(A)$

multiply both sides by $P(A)$ divide both sides by $P(B)$

$P(A \cap B) = P(A)P(B)$

The direction of all the thick arrows can also be reversed (provided one swaps roles between "multiply" and "divide"), which means that the two upper equalities imply each other, as indicated by the thin-line arrows. Independence is thus a two–way relation. One may justifiably say that "*A* and *B* are two independent events."

Note that the equality $P(A \cap B) = P(A)P(B)$ may serve as another equivalent definition of independence between *A* and *B*. It has the advantage of being symmetric with respect to *A* and *B*, and of not being limited to either $P(A) \neq 0$ or $P(B) \neq 0$, as are the other two definitions. It is, however, a less intuitive definition because in looking at it one cannot immediately see that the information that one of the events took place does not affect the probability of the other event.

b. To prove that positive dependence is symmetric, replace every equality sign in the above proof of **a** by a '$>$' sign.

c. To prove that negative dependence is symmetric, replace every equality sign in the above proof of **a** by a '$<$' sign.

2.3.16. Given $A \perp B$, we will use the equality

$$P(A \cap B) = P(A)P(B) \qquad (1)$$

to show that

$$P(\overline{A} \cap \overline{B}) = P(\overline{A})P(\overline{B}). \qquad (2)$$

Proof

$$P(\overline{A} \cap \overline{B}) = 1 - P(\overline{\overline{A} \cap \overline{B}}) = 1 - P(A \cup B). \qquad (3)$$

The second equality in (3) follows from De Morgan's law.

$$1 - P(A \cup B) = 1 - [P(A) + P(B) - P(A \cap B)]$$

$$= 1 - P(A) - P(B) + P(A \cap B) .$$

Now, replacing $P(A \cap B)$ by $P(A)P(B)$, as given in (1), we obtain

$$P(\overline{A} \cap \overline{B}) = 1 - P(A) - P(B) + P(A)P(B)$$

$$= [1 - P(A)][1 - P(B)] = P(\overline{A})P(\overline{B}) .$$

This completes the proof of (2).

2.3.18. Applying the formula of total probability,

$$P(W_2) = P(W_2 \mid W_1)P(W_1) + P(W_2 \mid B_1)P(B_1) ,$$

we obtain

$$P(W_2) = \frac{w - 1}{w + b - 1} \cdot \frac{w}{w + b} + \frac{w}{w + b - 1} \cdot \frac{b}{w + b}$$

$$= \frac{(w - 1)w + b\ w}{(w + b - 1)(w + b)} = \frac{(w - 1 + b)w}{(w + b - 1)(w + b)} .$$

Canceling out the nonzero factor $w + b - 1$, we obtain

$$P(W_2) = \frac{w}{w + b} \, .$$

Note: This proposition (and notation) may be extended to a larger number of successive draws without replacement from the same urn. That is, $P(W_i) = \frac{w}{w + b}$, for $i = 1, 2, \ldots, w+b$. (You may try to prove this as an extra challenge, or see the answer to Problem 2.5.12.)

2.3.19.   **a.** $P(W_2 \mid W_1) = 1/3$.
   b. $P(W_1 \mid W_2) = 1/3$.

Discussion. Section **b** of this problem usually triggers a heated discussion among students (Falk, 1979a, 1989). Some regard the problem as "meaningless" and refuse to consider it, claiming that conditioning the probability of an outcome of a draw on a subsequent event is not permissible.

Among those who do answer, the majority say that $P(W_1 \mid W_2) = 1/2$. These students typically argue that before the first draw the second draw has not yet been carried out, and "the first ball doesn't care whether the second is white or black." Therefore, they base their answer exclusively on the composition of the urn at the outset of the experiment, ignoring the information about the later outcome.

An appropriate response to such arguments is "indeed the first ball doesn't care whether the second is white or black but *we* do." The crux of the confusion is that the problem is addressed to our state of *knowledge*, rather than to the physical disposition of the urn to produce a certain result at a given moment (see Kahneman & Tversky, 1982). We have advanced beyond the initial stage when we learned that the second draw resulted in a white ball. This *information* removes one white ball out of the possible outcomes of the first draw, which are now one white and two blacks. Hence, $P(W_1 \mid W_2) = 1/3$, just like $P(W_2 \mid W_1)$.

An empirical verification of the solution to part **b** can be obtained by carrying out many repetitions of the experimental trial. Each trial is composed of two blind drawings without replacement from an urn containing two white and two black balls. We record the outcomes of the first and the second draw. Then we consider only those trials in which the second draw resulted in a white ball. Let $N(W_2)$ be the number of these trials. We count how many of them resulted also in white on the first draw. Denote that number $N(W_1 \cap W_2)$. We now compute the ratio $N(W_1 \cap W_2)/N(W_2)$. The greater the total number of trials, the closer that ratio will be to one third. (You may wish to try it yourself.)

A slight change in the experimental procedure may help eliminate any lingering doubts about the relevance of the information W_2 to the probability of W_1. Suppose we blindly draw a first ball from an urn containing two white and two black balls. Then, still blindly, we hide that ball in our pocket, and we go on and draw a second random ball. This time we look at the drawn ball and realize it is white (i.e., we observe W_2). What is *now* the probability that the ball in our pocket is also white, that is, what is $P(W_1 \mid W_2)$? Not only does this procedure make $P(W_1 \mid W_2)$ meaningful, it also helps us see why it should equal one third.

Note that $P(W_1) = P(W_2) = 1/2$ (see Problem 2.3.18). The result $P(W_2 \mid W_1) = 1/3$ can therefore be translated into $W_1 \searrow W_2$, as defined in Problem 2.3.15. The symmetry of negative dependence (part **c** of Problem 2.3.15), in turn, implies $W_2 \searrow W_1$, meaning that $P(W_1 \mid W_2)$ must be *less* than $P(W_1) = 1/2$. Thus, the answer $P(W_1 \mid W_2) = 1/2$ cannot be true since it would imply independence between W_1 and W_2, in contrast to the conclusion drawn from $P(W_2 \mid W_1) = 1/3$ (see also Borovcnik, 1988). Finally, since $P(W_1) = P(W_2)$, it follows from the definition of conditional probability that $P(W_1 \mid W_2)$ should equal $P(W_2 \mid W_1)$.

The lesson from the analysis of this problem should be that an event that occurs later than the target event can be relevant to our probabilistic inference and is perfectly legitimate as a conditioning event. The later event is informative, in the sense of changing our views, as long as there is statistical dependence (whether positive or negative) between the two events.

Evidence placed on the time axis later than the judged event may have *diagnostic* value. Psychologically, however, it is much easier to think in causal terms, in congruence with temporal order. Since causes come first and effects follow later, one tends to infer from causes to effects. Yet, we should realize that information about a later occurrence (or an effect) may have equally powerful impact on the probability that an earlier uncertain event (or a cause) has occurred. Instructive psychological investigations of causal schemas in judgments under uncertainty can be found in Tversky and Kahneman (1980).

Bayesian inference often relies on current observations to modify our probabilistic assessment of the *underlying situation*. This is routinely done in various areas of science and daily affairs as well. Thus, for example, the findings of archeological excavations shed new light on historical events which took place hundreds of years earlier. Recently obtained evidence is utilized to update police investigators' views on what 'really' took place in a murder case. In another context, although diseases are the causes and symptoms are the effects — although diseases come first and symptoms follow — the medical diagnostic procedure attaches probabilities to diseases on the basis of determination of their symptoms. Adjusting the probability of W_1, on the basis of observing W_2, is formally the same kind of inference.

2.3.20. Despite some intuitive appeal, this strategy *does not* increase the guesser's probability of winning so that it is higher than 1/36 (the probability of winning when randomly and independently picking numbers between 1 and 6 for the white and black dice).

There are six possible outcomes in which a white and a black die add to 7. If indeed the sum is 7, your probability of guessing the correct ordered pair is $1/6$. But, it is not certain that the sum will be 7; the probability of this event is $6/36 = 1/6$. Hence,

$$\begin{aligned}P(winning) &= P(sum\ 7)P(winning \mid sum\ 7) \\ &= (1/6) \times (1/6) = 1/36.\end{aligned}$$

2.3.21. The sex ratio of 1:1 will *not* change by implementing this policy. We leave it to the reader to figure out why.

2.4 Bayes' Theorem

2.4.1.

$$\begin{aligned}P(M \mid R) &= \frac{P(R \mid M)P(M)}{P(R \mid M)P(M) + P(R \mid F)P(F)} \\ &= \frac{0.95 \times 0.10}{0.95 \times 0.10 + 0.08 \times 0.90} = 0.57.\end{aligned}$$

2.4.2.   **a.** Let O denote the evidence that the suspect's blood type was found to be O. Given $P(O \mid G) = 1$ and $P(O \mid \overline{G}) = .33$, we are interested in the probability of guilt, in light of the evidence, namely, in $P(G \mid O)$. Now, employing Bayes' formula, we obtain

$$\begin{aligned}P(G \mid O) &= \frac{P(O \mid G)P(G)}{P(O \mid G)P(G) + P(O \mid \overline{G})P(\overline{G})} \\ &= \frac{1 \times .60}{1 \times .60 + .33 \times .40} = .82,\end{aligned}$$

verifying the intuition that the result of the blood test should be incriminating for the suspect, in the sense of raising his prior probability of guilt.

b. Common sense clearly indicates that blood type A would make the suspect *less* guilty than blood type O, since there are more people with blood type A. When substituting A for O as the evidence in Bayes' formula, the only change is replacing .33 in the denominator by .42, thus increasing the denominator and decreasing the resulting probability of guilt.

It is helpful to consider the limiting cases, where either 100% of the population shares the murderer's and suspect's blood type, leaving the suspect's prior probability of guilt unchanged, or where there is known to be just one person in the population with the blood type found to be the murderer's and the suspect's, causing our suspicion to jump to certainty. Altogether, the lower the rate of the blood type shared by the murderer and the suspect, the more incriminating the evidence.

2.4.3.   **a.** $P(G \mid E) = 0.60$.

   b. $P(G \mid E) = 1$.

   c. $P(G \mid E) = 0.75$.

   d. There is not sufficient information to determine $P(G \mid E)$. If, however, $P(E \mid \overline{G}) < 1$, then $P(G \mid E) > 0.50$.

2.4.4. **a.**

$$p = 1/2$$

Probability of Containing the Ship

Stage		Cell A	B	C
(0)	prior to any shooting	1/3	1/3	1/3
(1)	following "No" to cell A	1/5	2/5	2/5
(2)	following "No" to cell B	1/4	1/4	1/2
(3)	following "No" to cell C	1/3	1/3	1/3

b.

$$p = 2/3$$

Probability of Containing the Ship

Stage		Cell A	B	C
(0)	prior to any shooting	1/3	1/3	1/3
(1)	following "No" to cell A	1/7	3/7	3/7
(2)	following "No" to cell B	1/5	1/5	3/5
(3)	following "No" to cell C	1/3	1/3	1/3

2.4.5. **a.** $P_0(H) = 4/9$.
  b. $P_1(H) = 3/8$.
  c. $P_2(H) = 1$.

Derivation of solutions and discussion

a. Since both husband and wife have a B brother and D parents, these four parents must all be genetically bd. Consequently, the probability that the D husband is a carrier of a b gene is 2/3. The same is true for his wife. Because these two events are independent, the probability of their intersection is

$$P_0(H) = (2/3) \times (2/3) = 4/9; \text{ and } P_0(\overline{H}) = 5/9 \,.$$

Note that the focus of our discussion is an *unobservable* combination, or a *genotypic event*. The data used to infer the prior probability of that underlying state were *observables*, or phenotypes. Unlike situations in which the prior probability of the target event has to be subjectively assessed, in the present case that assessment is achieved by genetic inference based on information consisting of all the phenotypes and the relations in the pedigree.

b. The probabilities of obtaining D_1 under the two competing hypotheses are

$$P(D_1 \mid H) = 3/4 \text{ and } P(D_1 \mid \overline{H}) = 1.$$

Since the probability of having a dark-haired child is higher under $\overline{H}$ than under H, learning of the birth of a D child should somewhat *decrease* the probability of H. Indeed, the Bayesian computation yields such a decrease:

$$\begin{aligned}
P_1(H) &= P(H \mid D_1) \\
&= \frac{P(D_1 \mid H)P_0(H)}{P(D_1 \mid H)P_0(H) + P(D_1 \mid \overline{H})P_0(\overline{H})} \\
&= \frac{(3/4) \times (4/9)}{(3/4) \times (4/9) + 1 \times (5/9)} = \frac{3}{8}.
\end{aligned}$$

Some students find the direction of change of the target probability in this case intuitively obvious. It may, however, pose a difficulty to others who regard the information about the birth of a D child as worthless because it is inconclusive. They argue that this piece of evidence should not change the probability of H, since it does not rule out the possibility of both parents being heterozygous. At this point a thought-experiment in which the couple has 10 children, all dark haired, may help. It is easier to see that in this case the probability of H should be substantially reduced. It then follows that the birth of one D child should also have some impact on the probability. One lesson to be learned from this Bayesian analysis is that each item of information that is statistically dependent on the hypothesis in question is relevant. Thus, probabilistic inference via Bayes' theorem offers an extension of logical reasoning, as it allows our views to be revised even in light of inconclusive evidence. Although certainty is not achieved, our belief is modified .

c. The birth of a B child, however, is conclusive evidence. We do not need Bayes' formula to infer (with certainty) that the couple *is* capable of producing a B child. Nevertheless, it might increase our appreciation of the generality of Bayes' theorem to see that the formula yields the same conclusion:

$$
\begin{aligned}
P_2(H) &= P(H \mid B_2) \\
&= \frac{P(B_2 \mid H)P_1(H)}{P(B_2 \mid H)P_1(H) + P(B_2 \mid \overline{H})P_1(\overline{H})} \\
&= \frac{(1/4) \times (3/8)}{(1/4) \times (3/8) + 0 \times (5/8)} = 1.
\end{aligned}
$$

Thus we see that probabilistic inference in its extreme case (at the end of the probability scale) coincides with logical deduction.

The continuity of the process of probabilistic inference is well illustrated in the various parts of this problem by using the posterior probability of stage i, namely $P_i(H)$, as the prior of stage $i + 1$. In comparing the analyses of parts **b** and **c**, one may note that although $P(D_1 \mid \overline{H}) = 1$, observing D_1 (in **b**) did not prove $\overline{H}$, it only slightly increased its probability (from 5/9 to 5/8). This is compatible with intuition. For example, most people would know that even though rain implies clouds, clouds do not imply rain; they only make rain more probable. On the other hand, because $P(B_2 \mid \overline{H}) = 0$, observing B_2 (in **c**) eliminated $\overline{H}$ and proved H, fulfilling everybody's intuitive expectations.

This constitutes a demonstration of the conclusiveness of negative evidence compared with the relative corroboration affected by positive evidence.

2.4.6. This problem brings to mind a drunkard's slogan, contributed by Dave Allen to *Teaching Statistics* (1984), **6**, p.52. We offer this quote instead of an answer:

> Statistics show that 10% of all road accidents are caused by drunken drivers, which means that the other 90% are caused by sober drivers, ... so why can't they get off the road and leave us drunks to it?
>
> From M. Sandford

2.4.7. **a.** $P(S \mid +)$

b. $P_{30}(S \mid +) = 0.184$; $P_{45}(S \mid +) = 0.86\dot{5}$.

These results are more worrying for a 45-year-old woman.

2.4.8. **a.** 0.845

b. 0.215

c. 0.372

d. The overall probability of a correct diagnosis when using the test is 0.845 (in **a**). However, the probability of a correct diagnosis without using the test is 0.900, which is the base rate of normal people in this population (if we diagnose everybody as 'normal' we'll be right in 90% of the cases). This would seem to imply that the test does not improve our diagnostic ability, if our goal is to maximize correct diagnoses.

If our goal is to locate the sick and treat them, then without the test everybody would be treated, while only 10% would really need it (90% errors of futile treatments). Using the test — if only those who are diagnosed as psychotic are treated — 37.2% of those treated (see **c**) would in fact need it (the percentage of unnecessary treatments would be reduced to 62.8%). This seems to point in favor of using the test. At the same time, however, some of those diagnosed as normal may be psychotic (2.5%), and they would miss the chance of being helped.

Note that we may increase the rate of valid positives at the expense of a much lower rise in the false positive rate, and yet, because of the extremely skewed base rate (i.e., very low rate of the really sick persons in the population), that 'improvement' in the test's validity will result in an increase in the proportion of erroneous diagnoses (see Meehl, 1956). Thus, there is no simple answer to the question about a test's efficacy even when discussed only in probability terms. In addition to that, however, one should consider the *consequences* of the two types of possible errors and of correct diagnoses, weighing costs and benefits by their appropriate probabilities.

2.4.9. $1/5$

2.4.10. $5/13$

2.4.11. *Answer with discussion.* Both arguments make sense. We cannot decide between the two simply by discussing the problem, because each argument could be true depending on our assumptions. An explication of the assumptions, however, is missing in the problem's statement. The solution should depend on the exact method by which the observation has been obtained, namely, on the chance mechanism that has generated the datum. If the woman in question typically chooses at random one of her two children to accompany her on her outings, then your observation is "a randomly selected child from a two-child family was found to be a boy." In such a case, a family with two sons is twice as likely to yield such an observation than is a mixed family. The following simple Bayesian calculation shows that the posterior probability that the woman has two sons is $1/2$:

$$\frac{1 \times (1/4)}{1 \times (1/4) + (1/2) \times (1/4) + (1/2) \times (1/4) + 0 \times (1/4)} = \frac{1}{2}.$$

If, however, being a part of a male-chauvinistic society, Mrs. F. would always prefer taking a son along with her, then your observation becomes "the family has at least one son," and the three families of that kind are equally likely to lead to the above meeting. Consequently, the probability of two sons should be one third.

Textbooks of probability theory often begin with the concept of a *statistical experiment,* namely, an idealized, well-defined chance process whose elementary outcomes comprise the sample space (see, e.g., Feller, 1957, pp. 7–14). Indeed, probability theory is concerned with the outcomes of statistical experiments. As demonstrated in the analysis of the present problem, the probabilistic conclusions depend on the exact definition of the experiment's random process (see also Problem 2.2.12, answer and discussion). Describing the habits of the mother in choosing which child will accompany her to town amounts to modeling her behavior as formally equivalent to one urn model or another (Glickman, 1982). The statistical experiment should clarify whether one child is randomly sampled from a (random) two-child family, or whether one family is randomly sampled from two-child families having at least one son.

There is obviously no point in arguing about the correct answer to a problem that is under-defined. We could only *favor* one or the other of the assumptions as more reasonable. Lacking any other information, the most sensible assumption would be that chance determines which child accompanies the mother on a given trip. Indeed, in most real-life situations, you would learn that a given woman has a son in a way that is similar to this problem: you might overhear the woman mentioning a son, or you might pay a visit and catch a glimpse of a boy. In each of these cases, the greater the proportion of sons among the children in the family, the more likely is a chance meeting with one of them. Inventing a scenario in which such an encounter is equally likely under all three family compositions — BB, BG, and GB — is considerably more difficult (Falk, 1989; Loyer, 1983). The answer $1/2$ is thus favored over $1/3$ only in the sense that it is based on assumptions which we regard as more realistic.

2.4.12. *Discussion and detailed exposition of the answer.* Mrs. Montgomery was right in that after the first player survives pulling one of the crackers, the chance that a subsequent player will detonate the bomb when pulling the next cracker rises. On the other hand, she didn't know beforehand whether the first trial would turn out favorably. Were Mrs. Montgomery to blow up, the others' probability of exploding would drop to zero. The longer you wait, the lower your probability of survival, *provided that the bomb cracker is not pulled.* However, the longer you wait the more likely it is that someone else will pull the bomb before you.

The initial doubt about the very presence of the bomb seems to be enhanced when more and more crackers are safely pulled. But the lower number of remaining crackers, in its turn, tends to imply an increased risk. What do these antagonistic considerations amount to?

Interestingly, a careful quantitative analysis reveals that the contradicting tendencies balance out: *a priori, there is no advantage to any of the serial positions.* To prove this assertion let us first compute the posterior Bayesian probability for the presence of the bomb in the barrel, given that the first i crackers have been safely pulled.

Let B denote the event that *there is* a bomb in the barrel, and let b_i denote the event that player i explodes. P_i will designate the conditional probability that a bomb exists in the barrel, given that i crackers have been safely pulled. We were given the prior probability of the bomb's existence, $P_0 = 1/2$. By Bayes' theorem we obtain:

$$P_i = P(B \mid \bar{b}_1 \cap \bar{b}_2 \cap \ldots \cap \bar{b}_i) = \frac{\frac{6-i}{6} \times \frac{1}{2}}{\frac{6-i}{6} \times \frac{1}{2} + 1 \times \frac{1}{2}}$$

for $i = 1, 2, \ldots, 6$.

Altogether, the formula $P_i = \frac{6-i}{12-i}$ holds for $i = 0, 1, 2, \ldots, 6$.

The probability that the bomb is there *declines* as a function of the number of innocuous crackers that have been pulled. The course of this decreasing function is presented in the table on the next page (see also the Long-Term Hope Function, L_i, in the answer to Problem 2.4.13).

i	P_i
0	$\frac{6}{12} = \frac{1}{2}$
1	$\frac{5}{11}$
2	$\frac{4}{10} = \frac{2}{5}$
3	$\frac{3}{9} = \frac{1}{3}$
4	$\frac{2}{8} = \frac{1}{4}$
5	$\frac{1}{7}$
6	0

If the bomb is there, the first player will pull it with probability $1/6$. Since the prior probability of the bomb's existence in the barrel is $1/2$, the probability that the first player will blow up is $P(b_1) = (1/2) \times (1/6) = 1/12$.

We may now successively compute the probability of blowing up for each of the players (see Ayton & McClelland, 1987). For player i we need to multiply three factors: the probability that the previous $i - 1$ players did *not* pull the bomb; the conditional probability that the bomb is present, given that previous $i - 1$ players did not blow up (i.e., P_{i-1}); and the probability that player i pulls the bomb-cracker from the remaining $6 - i + 1$ crackers. These three factors are listed, in turn, in each row of the table below. Their product ends up giving each participant the same probability of pulling the bomb (i.e., $1/12$, see also the answer to Problem 2.3.3).

i	$P(b_i)$
1	$1 \times \frac{1}{2} \times \frac{1}{6} = \frac{1}{12}$
2	$\frac{11}{12} \times \frac{5}{11} \times \frac{1}{5} = \frac{1}{12}$
3	$\frac{10}{12} \times \frac{4}{10} \times \frac{1}{4} = \frac{1}{12}$
4	$\frac{9}{12} \times \frac{3}{9} \times \frac{1}{3} = \frac{1}{12}$
5	$\frac{8}{12} \times \frac{2}{8} \times \frac{1}{2} = \frac{1}{12}$
6	$\frac{7}{12} \times \frac{1}{7} \times 1 = \frac{1}{12}$

In hindsight, it should have been 'obvious' all along that (prior to starting the game) all the guests have the same chance of detonating the bomb. Suppose we introduce a slight change in the rules of the game. Instead of taking turns pulling a cracker, the six players are each assigned a cracker at random, and they pull them simultaneously. The mathematical structure of the problem has not changed. It is, however, easy to see that the modified version is symmetric with respect to all players. The pros and cons are identical for everybody and no player is favored over the others. (The same device can help us see the fairness of the procedure of drawing the matches in Problem 2.3.3).

2.4.13. **a.** (1) $\frac{7}{9} = 0.778$.

(2) $\frac{4}{6} = 0.667$.

(3) $\frac{1}{3} = 0.333$.

b. (1) $\frac{1}{9} = 0.111$.

(2) $\frac{1}{6} = 0.167$.

(3) $\frac{1}{3} = 0.333$.

Discussion. This problem concerns the course of your *'hope function'* throughout a search process. Part **a** addresses the *long-term* probability of finding the missing letter and part **b** the *short-term* probability. The required probabilities are conditioned, in both cases, on the information that you did not find the letter in the first i drawers.

The Long-Term Hope Function is defined as

$$L_i = P(\text{letter in desk} \mid \text{letter wasn't in 1st } i \text{ drawers}),$$

for $i = 0, 1, 2, \ldots, 7, 8$.

The Short-Term Hope Function is defined as

$$S_i = P(\text{ letter in next drawer} \mid \text{letter wasn't in 1st } i \text{ drawers}),$$

for $i = 0, 1, 2, \ldots, 7$.

The course of these conditional probabilities, as functions of the ongoing failure to find the letter, is *not* intuitively clear to most people, as established by Falk, Lipson, and Konold's (in press) exploratory study. The six numerical answers (listed above) can be obtained by inserting the given numbers into Bayes' theorem. In order to make the solutions more intuitively acceptable, however, we consider a slight change[7] in the formulation of the problems, as presented below.

Without loss of generality, let us suppose that there are *ten* drawers in your desk. Your assistant *always* puts your letters in the drawers of your desk after you have read them. You know that the letter is *equally likely* to be in any of the ten drawers. You notice, however, that drawers # 9 and # 10 are *locked* (see figure), and your assistant has gone home with the keys. You realize that there is an 80% chance that the letter is in one of the unlocked drawers. You start a thorough and systematic search of the eight unlocked drawers.

---

[7]Suggested by Oren Falk and by Cliff Konold

1	o
2	o
3	o
4	o
5	o
6	o
7	o
8	o
9	Locked
10	Locked

All the questions are exactly as before. In **a** you are asked three times about the (conditional) probability that the letter is *in one of the unlocked drawers*, and in **b** about the (conditional) probability that it is *in the next drawer*.

Now, considering the new (isomorphic) version of the problems, you may obtain the answers given above by simple proportional reasoning.[8] Moreover, you can easily derive the respective formulas for the functions L_i and S_i, required in parts **a** and **b**:

a. $L_i = \dfrac{8-i}{10-i}$ for $i = 0, 1, 2, \ldots, 8$.

b. $S_i = \dfrac{1}{10-i}$ for $i = 0, 1, 2, \ldots, 7$.

It is noteworthy that the long-term hope of finding the letter decreases as more and more drawers are found not to contain the letter (as is the case with Doctor Fischer's bomb in Problem 2.4.12), whereas the immediate hope of finding the

---

[8]This obviously requires a careful consideration of the *changing base* for computing the required proportion in each stage. As rightly emphasized by Hooke (1983), in a chapter headed *The Unmentioned Base*: "Every percentage is a percentage *of* something, and this something is called the base. Since the base is often understood and not mentioned, changing it to cause confusion is easy to do" (p. 14).

letter in the next drawer increases throughout in the process (see the figure below). The latter increase in the short-term conditional probability is analogous to the mounting risk of blowing up at Doctor Fischer's bomb party, as more and more participants survive pulling the crackers (see Problem 2.4.12).

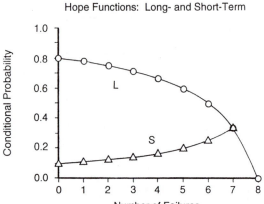

Hope Functions: Long- and Short-Term

2.4.14. *Solution and discussion.* If we denote the events that the car is behind door No. 1, 2, or 3 by A, B, and C, respectively, it is clear that *a priori*: $P(A) = P(B) = P(C) = 1/3$. By the rules of this game, Monty is supposed to open a door that the contestant didn't choose, which has a goat behind it. Thus, given that the contestant picks door No. 1, if the car is behind door No. 2 Monty will open No. 3, and if it is behind No. 3 he'll open No. 2.

Let b denote the observation that Monty opens door No. 2 to reveal a goat, and c that he does so with door No. 3. We know that $P(b \mid B) = 0$; $P(b \mid C) = 1$, and obviously $P(c \mid B) = 1$; $P(c \mid C) = 0$. If, however, the car is behind door No. 1 (the contestant's original choice), Monty can open either door No. 2 or door No. 3. How does Monty behave when confronted with this choice? The problem's text provides no clues. In this sense, the description of the 'statistical experiment' that has generated our observation is incomplete. The solvers must supply their own assumptions. As we saw (see answers to Problems 2.2.12 and 2.4.11), different random procedures which yield the same data may result in different answers.

The first assumption that comes to mind is that Monty is *indifferent* as to whether he opens door No. 2 or No. 3. If he could open either, he'll (secretly) flip a coin to determine which door to open. This seems the fairest assumption in the absence of information about Monty's preferences. This assumption, put in symbols, is: $P(b \mid A) = P(c \mid A) = 1/2$. Given that Monty has opened door No. 3 (i.e., c has been observed and C has been ruled out) we can now compute the posterior probability that the car is behind door No. 1:

$$
\begin{aligned}
P(A \mid c) &= \frac{P(c \mid A)P(A)}{P(c \mid A)P(A) + P(c \mid B)P(B)} \\
&= \frac{\frac{1}{2} \times \frac{1}{3}}{\frac{1}{2} \times \frac{1}{3} + 1 \times \frac{1}{3}} = \frac{1}{3}.
\end{aligned}
$$

It follows that $P(B \mid c) = 2/3$ and $P(C \mid c) = 0$, and the contestant's conclusion should be to *switch* to door No. 2.

The heated public discussion that grew around that problem can be essentially accounted for by a prevalent tendency to attribute *equal probabilities* to the available options. Many people insist that having door No. 3 ruled out, the remaining two doors are equally likely to hide the car and, consequently, there is no reason to switch. They seem to jump into the uniformity conclusion confidently and spontaneously. They do not even question the validity of that belief.

In her "Ask Marilyn" column in *Parade Magazine*, Marilyn vos Savant reported receiving thousands of letters subsequent to her publishing the solution advocating a switch to door No. 2. Many of the letters were from highly educated university people. About 90% of them smugly reprimanded her for being in error (*Parade Magazine*, December 2, 1990, p. 25 and February 17, 1991, p. 12). Her 'error' was in departing from the equiprobability of the remaining possibilities. Here are two sample letters:

1. If one door is shown to be a loser, that information changes the probability to 1/2. As a professional mathematician, I'm very concerned with the general public's lack of mathematical skills. Please help by confessing your error and, in the future, being more careful.

2. You blew it, and you blew it big! I'll explain: After the host reveals a goat, you now have a one-in-two chance of being correct. Whether you change your answer or not, the odds are the same. There is enough mathematical illiteracy in this country, and we don't need the world's highest IQ [referring to Marilyn, the column's editor] propagating more. Shame!

Obviously, the uniformity belief is deeply ingrained in many minds. The primacy of the uniformity intuition can be traced historically to early stages in the development of probability theory (Gigerenzer, et al., 1989; Hacking, 1975, chap. 14; Zabell, 1988c). Although the justifiability of the assumption is, and has always been, subject to profound doubts and debates, there is no doubt about the supremacy of the psychological proclivity toward the uniformity belief. Dividing the uncertainty equally among the existing possibilities is apparently induced by a basic preference for *symmetry* and *fairness*, as reflected in expressions such as "insufficient reason," "equal ignorance," and the "principle of indifference." After observing a goat behind door No. 3, however, there is no reason to presume "equal ignorance" anymore. That evidence should have differential impact on the chances of the two remaining doors to hide the desired prize. Indeed, it is often the case in Bayesian analysis that initially equally-likely events assume unequal probabilities when revised in the light of new observations (Bar-Hillel & Falk, 1982).

At the risk of being repetitious, we stress again the dependence of the Bayesian result — assigning a probability of 2/3 to winning the prize by switching to door No. 2 — on the assumption that Monty is unbiased about opening either door No. 2 or No. 3 when he knows the prize to be behind No. 1. Otherwise, if, for example, Monty is committed (for some idiosyncratic reason of his own) to opening door No. 3 whenever allowed, and you know about that bias, it would mean $P(c \mid A) = 1$. In such a case, observing a goat when door No. 3 is opened would result in equal posterior probabilities that doors No. 1 and 2 hide the prize (the reader is urged to verify this Bayesian result), and there would be no reason to switch.

Since you don't know whether Monty is biased, to start with, and if he is, you don't know in favor of which door and to what extent, switching is always recommended. Let us assume that Monty will open door No. 3 with probability p (where $0 \leq p \leq 1$) when the prize is behind door No. 1. By Bayes' theorem, we find that the posterior probability P that door No. 2 hides the prize, after observing a goat behind door No. 3, is $P = \frac{1}{1+p}$. This means that the probability of winning upon switching satisfies $1/2 \leq P \leq 1$. It would therefore be wise for the contestant to switch (Gillman, 1991, 1992).

Monty's problem, just like another well-known isomorphic problem, the problem of the three prisoners,[9] brings to the fore people's intuitive beliefs. Another widespread belief, which is incompatible with the uniformity assumption, is the *no-news-no-change belief*: since you know from the beginning that Monty intends to open one of the doors you don't choose to reveal a goat, and since he can always do it, observing the goat behind door No. 3 provides no new information and should induce no change in your views. At face value, there is something very compelling about the idea that when we receive a piece of information we have known all along, it should not alter our assessment of the situation. This view is reinforced

---

[9]See, for example, Diaconis and Zabell (1986), Falk (1992), Gardner (1961, 1992), Mosteller (1965), Shimojo and Ichikawa (1989), Székely (1986), and Zabell (1988b).

by the Bayesian calculation showing that when incorporating
the observation of the goat behind door No. 3, the probability
of the car being behind door No. 1 still stays unchanged (it is
also expressed by Marilyn in *Parade Magazine*, December 2,
1990, p. 25, and by Gardner, 1992). It is nevertheless at fault.

Truly, Monty can always open one of the two other doors to
show a goat, *and* the probability of door No. 1 remains unchanged subsequent to observing that goat, still, it is *not because* of the former that the latter is true. The probability of
winning the car by sticking with door No. 1 remains unchanged
due to a specific combination of priors and likelihoods characterizing this problem. Had Monty been biased in favor of
opening door No. 3, it would still be true that he can always
open one of the other two doors to reveal a goat, and yet, observing that goat would change your prior probability of 1/3
for door No. 1 to 1/2.

In addition, if we postulate *unequal priors* for the three doors,
such as $P(A) = 1/4$, $P(B) = 1/2$, $P(C) = 1/4$, and keep
the assumption that Monty flips a coin to decide which door
to open when confronted with a choice, it would certainly be
true that he can always open one of doors No. 2 or No. 3 to
show a goat, still the observation of a goat behind door No. 3
would *change* the prior of No. 1. The result (to be verified by
the reader) is $P(A \mid c) = 1/5$ (see Shimojo & Ichikawa, 1989).

We conclude by reiterating the crucial role of a proper definition of the statistical experiment. Marilyn happened to forget
this for a moment when (trying to convince her stubborn cor-
respondents of the need to switch) she resorted to an extreme
version of the problem:

> Suppose there are a *million* doors, and you pick
> door No. 1. Then the host, who knows what's be-
> hind the doors and will always avoid the one with
> the prize, opens them all except door No. 777,777.
> You'd switch to that door pretty fast, wouldn't you?
> (*Parade Magazine*, February 17, 1991, p. 12).

There can hardly be a more smashing argument. Except, the argument's truth depends on the understanding that if the prize is behind door No. 1, the host decides by a fair draw which one of the remaining 999,999 doors to leave closed. If, however, you know that for some reason or other the host is determined to leave door No. 777,777 closed, whenever possible, observing that situation will render that door as likely to hide the prize as door No. 1. You would then have no reason to hurry to switch.

2.5 Probability Distributions and Expectations

2.5.1. **a.** $P(X = i) = \frac{1}{4}(\frac{3}{4})^{i-1}$, for $i \geq 1$ (Compare with Problem 2.5.5.a).

       b. $P(X = i) = \frac{1}{4}$, for $i = 1, 2, 3, 4$.

       c. $P(X = 1) = \frac{1}{4}$; $P(X = i) = \frac{1}{4}(\frac{2}{3})^{i-2}$, for $i \geq 2$.

2.5.2. **a.** Denote the amount (in dollars) that he would be willing to pay by x; then $0.75 \times 12 - 0.25x = 2$. The solution is $x = \$28$.

       b. Zero

2.5.3. $\$0.128$.   Obtained by the computation $0.09 \times 0.05\sum_{i=1}^{8} i0.95^{i-1}$.

2.5.4.   **a.**

k	$P(X = k)$
0	$\frac{4}{7} = 0.571$
1	$\frac{3}{7} \times \frac{4}{6} = 0.286$
2	$\frac{3}{7} \times \frac{2}{6} \times \frac{4}{5} = 0.114$
3	$\frac{3}{7} \times \frac{2}{6} \times \frac{1}{5} = 0.029$
Total	1

b. $E(X) = 0.601$.

2.5.5.   **a.** $P(X = k) = \left(\frac{3}{7}\right)^k \frac{4}{7}$, for $k = 0, 1, 2, \ldots$. This is an example of the so-called *geometric distribution*. Unlike the case in Problem 2.5.4.**a**, the random variable X in this case is infinite. (Compare with Problem 2.5.1**a**)

b. $E(X) = 0.750$.

The computation of this expected value involves some manipulations of infinitely-decreasing geometric progressions. (See also Problem 2.5.6**b**)

2.5.6.   **a.** $\frac{1}{n}\sum_{i=1}^{n} i = \frac{n+1}{2}$.

b. n (Compare with Problem 2.5.5**b**)

2.5.7.   **a.** $\left(\frac{1}{6}\right)^3 + 3 \times \left(\frac{1}{6}\right)^2 \times \frac{5}{6} + 3 \times \frac{1}{6} \times \left(\frac{5}{6}\right)^2 = 0.421$.

Can you find a shorter (but indirect) way of computing this probability?

b. $3 \times \left(\frac{1}{6}\right)^3 + 2 \times 3 \times \left(\frac{1}{6}\right)^2 \times \frac{5}{6} + 3 \times \frac{1}{6} \times \left(\frac{5}{6}\right)^2 - \left(\frac{5}{6}\right)^3 = -0.0787$.

There is an expected *loss* of $0.079 in playing chuck-a-luck.

c. $20

2.5.8. The table below presents the payoffs (in dollars) for each possible case:

Family's decision	Fate of luggage	
	Is lost (probability 0.01)	Arrives safely (probability 0.99)
Buy insurance	$3000 - (2500 + 150) =$ $= 350$	-150
Do not buy insurance	-2500	0

a.

$$0.01(-2500) + 0.99 \times 0 = -25.$$

The expected 'gain' without insurance is -$25.

b.

$$0.01 \times 350 - 0.99 \times 150 = -145.$$

The expected 'gain' if the family buys insurance is $-$145.

From a strict monetary point of view, buying the insurance incurs a greater expected loss than sending the luggage uninsured. Yet, evaluating this decision is not simple. Buying insurance may be quite rational if one considers nonmonetary assets such as peace of mind. In addition, the (subjective) value of money for people ('utility') might be a nonlinear function of the number of dollars, so that the expected utility of buying insurance could be positive.

2.5.9.

$$1 \times 0.99^{50} + 51(1 - 0.99^{50}) = 20.7\bar{5} \sim 21.$$

The expected number of tests upon pooling 50 blood samples is approximately 21. Compared with 50 individual tests, it will indeed save labor in the long run.

2.5.10.   The Crude Death Rate (CDR) is a global measure of mortality. CDR is determined not only by all the Age-Specific Death Rates (ASDR's), but also by the *age distribution* of the population. The CDR is, in fact, the *weighted mean* of all ASDR's with the relative frequencies of the different age groups serving as weights.

Because the Jewish subpopulation is 'older' than the Arabic subpopulation, the ASDR's for the older age groups are weighted more heavily when computing the CDR for the Jewish subpopulation. Within the Jewish subgroup, the ASDR's for the older ages are higher than those of other age groups.

Heavier weights for relatively greater components within the Jewish subpopulation result in a greater weighted average than that of the Arabs, although each averaged ASDR in the Arabic subpopulation is greater than the corresponding ASDR in the Jewish subpopulation. (A similar intriguing situation is presented in Problem 2.3.10.)

Note that because the Jewish subpopulation is older (i.e., older age groups are relatively more probable among the Jews than among the Arabs) but its age-specific death rates are lower, both its CDR and its *Life Expectancy* (a measure which is, in a way, the inverse of mortality) are higher than those of Israeli Arabs.

2.5.11. [10]

    a. Let $p = N_1/N$ denote the success probability in (randomly) sampling one of the N elements. The probability of a 'failure' will be denoted $q = 1 - p$. The probability distribution of X is given by

$$P(X = k) = \binom{n}{k} p^k q^{n-k}, \qquad k = 0, 1, 2, \dots, n. \qquad (1)$$

This is the so-called *binomial distribution*, that is, the distribution of 'number of successes' in n Bernoulli trials. (See Feller, 1957, p. 137; Hays & Winkler, 1971, pp. 181–185; Hogg & Tanis, 1977, pp. 66–68; and Walpole, 1974, pp. 79–81.)

    b.

$$P(Y = k) = \frac{\binom{N_1}{k}\binom{N - N_1}{n - k}}{\binom{N}{n}}, \qquad k = 0, 1, \dots, n. \quad (2)$$

This is the so-called *hypergeometric distribution*. (See Feller, 1957, pp. 41–42; Hays & Winkler, 1971, pp. 201–202; Hogg & Tanis, 1977, p. 26; and Walpole, 1974, pp. 86–88.)

    c. Formulas (1) and (2) seem quite dissimilar. This is partly because the probabilities p and q are used in (1) and the absolute frequencies N_1 and $N - N_1$ are used in (2). If we substitute N_1/N for p and $(N - N_1)/N$ for q in (1), we obtain, after some minimal algebraic manipulations, a formula *equivalent* to the binomial distribution (1):

$$P(X = k) = \binom{n}{k} \frac{N_1^k (N - N_1)^{n-k}}{N^n}. \qquad (1')$$

Likewise, a few algebraic manipulations yield a formula *equivalent* to the hypergeometric distribution (2):

$$P(Y = k) = \binom{n}{k} \frac{(N_1)_k (N - N_1)_{n-k}}{(N)_n}. \qquad (2')$$

---

[10]See Falk (1986c).

Note the similarity in the structures of formulas (1′) and
(2′). The only dissimilarity is that whenever an exponential of the form s^t appears in the binomial formula,
(1′), the term $(s)_t$ appears instead in the hypergeometric
formula (2′). (See Problem 2.2.12.) Both s^t and $(s)_t$ are
products of t factors:

$$s^t = \underbrace{s \, s \, \ldots \, s}_{t \text{ factors}}$$

$$(s)_t = \underbrace{s \, (s-1) \, \ldots \, (s-t+1)}_{t \text{ factors}}.$$

However, in s^t all the factors equal s, reflecting the independence of the different samplings (with replacement),
whereas in $(s)_t$ the factor s is reduced by 1 in each successive step since one element is removed from the population with each act of sampling (without replacement).

2.5.12. To simplify matters, we denote the probability of 'success' in
one random draw by $p = N_1/N$, and the probability of 'failure'
by q (obviously, $p + q = 1$). When sampling *with* replacement
the probability of success on the ith draw is clearly p, for
$i = 1, 2, \ldots, n$. This is, however, true for sampling *without*
replacement as well.

That the probability of success on the ith draw is p also when
the sampling is conducted without replacement, may not be
immediately obvious. One has to remember, however, that we
are not seeking the conditional probability of a success on the
ith draw, given a specific subsequence of previous outcomes,
but rather the total probability, across all possible outcomes
of the previous $i - 1$ draws (see Problem 2.3.18). Some authors (such as Hays & Winkler, 1971) take this equality for
granted, others justify it by the phrase "for reasons of symmetry" (Feller, 1957, p. 218). The reader who still has doubts

about this assertion should note that the total number of possible orderings of a population of N elements, divided into N_1 successes and $N - N_1$ failures, is $\begin{pmatrix} N \\ N_1 \end{pmatrix}$. (See answer to Problem 2.2.5.c and answer to Problem 2.2.12.) The number of 'favorable' orderings (i.e., those resulting in success on the ith place) is $\begin{pmatrix} N - 1 \\ N_1 - 1 \end{pmatrix}$. Dividing the second number by the first yields $N_1/N = p$.

The computations of the expected values of the binomial and the hypergeometric random variables can now be derived together. Consider n indicator variables, $I_1, I_2, \ldots, I_n$, where I_j assumes a value 1 or 0, depending on whether we have a success or a failure on the jth draw. Both random variables — the binomial, X, and the hypergeometric, Y — can be written as a sum of these n variables. The only difference is that the n variables are independent when sampling with replacement (binomial) and dependent when sampling without replacement (hypergeometric). A basic statistical theorem states that given some finite number of random variables, the expectation of the sum of those variables is the sum of their individual expectations. This theorem is not qualified by any requirement concerning the dependence relations between the variables. It is valid for both dependent and indpendent variables (see also Problems 2.5.13 and 2.5.15). Therefore, once we derive the expected value of each indicator variable,

$$E(I_j) = 1 \cdot p + 0 \cdot q = p \qquad j = 1, 2, \ldots, n \, ,$$

we easily obtain the expected value[11] of the binomial as well as that of the hypergeometric random variable,

$$E(I_1 + I_2 + \ldots + I_n) = E(I_1) + E(I_2) + \ldots + E(I_n) = np ,$$

that is, $E(X) = E(Y) = np = n\frac{N_1}{N}$.

2.5.13.   **a.** A straightforward method (feasible for $n = 4$) of figuring out the expected value is to list all 24 permutations x_1, x_2, x_3, x_4 of 1, 2, 3, 4 with the corresponding payments in dollars (Freudenthal, 1970, p. 162):

1 2 3 4 ; **4**	2 1 3 4 ; **2**	3 1 2 4 ; **1**	4 1 2 3 ; **0**
1 2 4 3 ; **2**	2 1 4 3 ; **0**	3 1 4 2 ; **0**	4 1 3 2 ; **1**
1 3 2 4 ; **2**	2 3 1 4 ; **1**	3 2 1 4 ; **2**	4 2 1 3 ; **1**
1 3 4 2 ; **1**	2 3 4 1 ; **0**	3 2 4 1 ; **1**	4 2 3 1 ; **2**
1 4 2 3 ; **1**	2 4 1 3 ; **0**	3 4 1 2 ; **0**	4 3 1 2 ; **0**
1 4 3 2 ; **2**	2 4 3 1 ; **1**	3 4 2 1 ; **0**	4 3 2 1 ; **0**

Note that there are no payments of **3** (make sure you see why). The sum of the payments is **24**, *the expectation is 1 dollar.*

An alternative way to find the same expectation is to use the addition law of expectations. This method spares us from enumerating all permutations, which becomes extremely cumbersome as n grows beyond 4. The addition law states that the expectation of the sum of random variables equals the sum of their expectations (we consider

---

[11]The variances of the binomial and the hypergeometric distributions are not equal anymore. The variance of a sum of random variables equals the sum of the individual variances, provided every two variables are uncorrelated. This condition holds in the binomial but not in the hypergeometric case. The direction of the difference between the variances of the two distributions can be anticipated: the hypergeometric variance should be smaller than the binomial variance. This is so because the covariances between pairs of the above indicator variables are negative when sampling without replacement, since, once a certain draw results in a success, the probability of success on another draw is reduced (See Feller, 1957, pp. 218–219 for the exact derivation of the variances of the two distributions).

here only the case of a finite number of variables). This theorem provides a powerful tool for solving problems like the present one because it holds irrespective of whether the variables in question are dependent or independent (they are dependent in our case, see also Problem 2.5.12 and Problem 2.5.15).

Let Y_i be the payment received for the ith number drawn from the bag (denoted x_i). It is defined by

$$Y_i = \left\{ \begin{array}{ll} 0 & \text{if } x_i \neq i \\ 1 & \text{if } x_i = i \end{array} \right\} \quad \text{for } i = 1, 2, 3, 4.$$

It can be seen that $P(Y_i = 1) = \frac{1}{4}$, because there are always six permutations of the 24 in which a given number is placed in the correct box. It follows that $E(Y_i) = \frac{1}{4}$, for $i = 1, 2, 3, 4$.

Now, applying $E(\sum_{i=1}^{4} Y_i) = \sum_{i=1}^{4} E(Y_i)$, we obtain the expectation of the sum of the Y_i's, that is, the player's expected total payment: $E(\sum_{i=1}^{4} Y_i) = 4 \times \frac{1}{4} = 1$.

b.

k	$P(Y = k)$
0	$\frac{9}{24}$
1	$\frac{8}{24}$
2	$\frac{6}{24}$
3	0
4	$\frac{1}{24}$
Total	1

$$E(Y) = \frac{9}{24} \times 0 + \frac{8}{24} \times 1 + \frac{6}{24} \times 2 + 0 \times 3 + \frac{1}{24} \times 4 = \frac{24}{24} = 1,$$

confirming the result in **a**. Note that the random variable Y equals $\sum_{i=1}^{4} Y_i$, where the Y_i's are defined in **a**.

c. In the general case,[12] let Y_i be defined by

$$Y_i = \left\{ \begin{array}{ll} 0 & \text{if } x_i \neq i \\ 1 & \text{if } x_i = i \end{array} \right\} \qquad \text{for } i = 1, 2, 3, \ldots, n.$$

The probability of a match in the ith place is $\dfrac{(n-1)!}{n!} = \frac{1}{n}$, which means that $P(Y_i = 1) = \frac{1}{n}$, and therefore $E(Y_i) = \frac{1}{n}$, for $i = 1, 2, 3, \ldots, n$. The expected number of matches (the expected payment for the game) is thus,

$$E(\sum_{i=1}^{n} Y_i) = \sum_{i=1}^{n} E(Y_i) = n \times \frac{1}{n} = 1 .$$

The expectation of the number of matches *does not* depend on n; it is always 1 (some people find this result unintuitive).

2.5.14. [13] There are altogether $\begin{pmatrix} N_1 + N_2 \\ N_1 \end{pmatrix}$ permutations of a collection of N_1 indistinguishable elements of one type and N_2 of another type. (See Problem 2.2.5.**c**, answer to this problem, and illustration on p. 148).

To find $P(R = r)$ for an *even* number r, we note that there are $\frac{r}{2}$ runs of X's and $\frac{r}{2}$ runs of O's. The sequence can either start with a run of X's and end with one of O's or vice versa. The N_1 X's can be divided into $\frac{r}{2}$ runs by inserting runs of O's in any $\frac{r}{2} - 1$ places out of the $N_1 - 1$ transitions between successive X's. Likewise, in order to get $\frac{r}{2}$ runs of O's, we must break the succession of N_2 O's in any $\frac{r}{2} - 1$ places out of the $N_2 - 1$ transitions between consecutive O's. We can then,

---

[12]See Feller (1957, p. 217)

[13]See Feller (1957, p. 60), Hays & Winkler (1971, p. 826), Hogg & Tanis (1977, pp. 277–278), and Siegel (1956, p. 138).

either starting with the X's or the O's, create a sequence with $R = r$ by alternating runs of X's and O's. These considerations result in

$$P(R = r) = \frac{2 \left(\begin{array}{c} N_1 - 1 \\ \frac{r}{2} - 1 \end{array} \right) \left(\begin{array}{c} N_2 - 1 \\ \frac{r}{2} - 1 \end{array} \right)}{\left(\begin{array}{c} N_1 + N_2 \\ N_1 \end{array} \right)}.$$

For an *odd* number r, such that $r = 2k + 1$, the combinatorial considerations are quite similar. The only difference is that there are either k runs of X's and $k + 1$ runs of O's (as is the case in the sequence shown in the statement of Problem 2.5.14, where $k = 2$), or vice versa. These considerations result in

$$P(R = r) =$$

$$\frac{\left(\begin{array}{c} N_1 - 1 \\ k - 1 \end{array} \right) \left(\begin{array}{c} N_2 - 1 \\ k \end{array} \right) + \left(\begin{array}{c} N_1 - 1 \\ k \end{array} \right) \left(\begin{array}{c} N_2 - 1 \\ k - 1 \end{array} \right)}{\left(\begin{array}{c} N_1 + N_2 \\ N_1 \end{array} \right)}.$$

Note: See Problem 2.5.15 for the expectation of the number of runs in sequences of this type.

2.5.15. To compute $E(A)$, we define $N - 1$ variables $X_1, X_2, \ldots, X_{N-1}$, as follows: X_i equals 1 when symbols number i and $i + 1$ are different, and 0 when there is no change of symbols (continuity) on the ith transition. This definition holds for $i = 1, 2, \ldots, N - 1$. We have $A = X_1 + X_2 + \ldots + X_{N-1}$.

For an alternation to occur in, say, the first transition, either the first symbol should be X and the second one O or vice versa, and these two events are mutually exclusive. Hence,

$$P(X_1 = 1) = \frac{N_1}{N} \frac{N_2}{N - 1} + \frac{N_2}{N} \frac{N_1}{N - 1} = 2 \frac{N_1 N_2}{N(N - 1)}.$$

For reasons of symmetry,[14] this equals the probability of alternation for *any transition* in the sequence, namely,

$$P(X_i = 1) = 2\frac{N_1 N_2}{N(N-1)}; \qquad i = 1, 2, \ldots, N-1.$$

Since the other value that X_i may assume is 0, we obtain

$$E(X_i) = 2\frac{N_1 N_2}{N(N-1)}; \qquad i = 1, 2, \ldots, N-1.$$

Applying the addition law of expectations, we obtain

$$E(A) = 2\frac{N_1 N_2}{N},$$

and consequently

$$E(R) = 2\frac{N_1 N_2}{N} + 1.$$

This result agrees with the formula that is usually given without proof in textbooks of statistics (see, e. g., Hogg & Tanis, 1977, p. 278; Siegel, 1956, p. 56).

---

[14]A direct computation of the probability of alternation on the ith transition can be worked out as well:

The total number of possible sequences of N_1 X's and N_2 O's is $\begin{pmatrix} N_1 + N_2 \\ N_1 \end{pmatrix}$.
All of these sequences are equiprobable if arranged at random. For an alternation to occur on the ith transition, one should either have an O in the ith place and an X in the $(i+1)$th place, or vice versa. In both cases, the possible number of arrangements of the remaining $N_1 + N_2 - 2$ symbols over the remaining places is $\begin{pmatrix} N_1 + N_2 - 2 \\ N_1 - 1 \end{pmatrix}$, which is half the number of favorable outcomes. The probability of alternation is therefore

$$P(X_i = 1) = 2\frac{\begin{pmatrix} N_1 + N_2 - 2 \\ N_1 - 1 \end{pmatrix}}{\begin{pmatrix} N_1 + N_2 \\ N_1 \end{pmatrix}}.$$

Upon expanding the binomial coefficients, some factors can be canceled. Replacing $N_1 + N_2$ by N, we obtain

$$P(X_i = 1) = 2\frac{N_1 N_2}{N(N-1)} \qquad \text{for } i = 1, 2, \ldots, N-1.$$

The derivation of the formula for $E(R)$ provides another demonstration of the efficacy of the law stating that the expectation of a sum of random variables equals the sum of the expectations, which holds whether the added variables are independent or dependent (they are dependent in this problem). See also Problems 2.5.12 and 2.5.13.

2.5.16.   **a.**

	Game 1	Game 2
Shape of the distribution	Approximately normal	Uniform (discrete)
Expected value	$50 \times 3.5 = 175$	$50 \times 3.5 = 175$
Variance	2.9167×50 $=145.8$	2.9167×50^2 $=7291.7$
Standard deviation	12.076	85.391

   b. In Game 1: the probability is $1/6^{50}$, which is practically zero.

In Game 2: the probability is $1/6$.

2.5.17. To verify whether the argument is valid for *any positive A* that may be shown to you, we should examine what we know about the range and distribution of the positive numbers written on the two cards. How did these numbers come about? We need to know the chance mechanism that has generated them. Nothing, however, is said about the selection process in the problem.

Suppose you were told that a person had drawn at random one of the integers 1, 2, 3, ..., 10; the outcome of that draw was written on one card, and twice that number on the other card. Now, considering this experimental procedure, if the number revealed to you were, say, 18, you would know that you should stick with it. If, however, you were shown a 7, you would switch. Different considerations apply, depending on the number revealed to you and on the description of the number-generating procedure.

Given the description of the statistical experiment by which the two numbers have been chosen, you can assess the prior distribution of the pairs of values. Once you observe one of the numbers, you can apply Bayesian computation to find the posterior expectation of winnings conditioned on your switching. Lacking any specification of the chance set-up in the problem's statement, the computation must be based on some *assumed* distribution of the pairs of numbers.

The crux of the puzzle is based on the assertion that given any $A > 0$, the probabilities that the other card bears either $0.5A$ or $2A$ are equal. The one assumption that is compatible with this requirement for *each* positive A, at least in the discrete case, is that the first number was *randomly* drawn from the set of, say, all integer powers of 2, and then it was doubled. Put differently, your expected winnings upon switching would be $1.25A$, *for every* A, only if you assume that the first number was randomly drawn from an infinite uniform distribution.[15]

A uniform distribution, however, *cannot* be infinite. If a discrete random variable is uniformly distributed, it can assume only a finite number of values. Both infinity and uniformity, however, had been tacitly assumed in the exposition of the above problem. On first encounter, the reader is not aware of the fact that he or she is accepting an impossible statistical experiment. The paradoxical conclusion can, therefore, be traced to these contradictory implicit assumptions (see also Comments on "Where the Grass is Greener" in the *College Mathematics Journal* of September 1991, Vol. **22**, No. 4, pp. 308–309). A related puzzle ("The Wallet Game") can be found in Gardner (1982, p. 106).

---

[15] We are grateful to Ester Samuel-Cahn for her illuminating comments concerning the resolution of this paradox.

Is there an underlying statistical experiment according to which it is always reasonable to switch? Suppose the game is conducted so that you are first shown some positive number W. Then, out of your sight, somebody flips a coin to decide whether the other card will bear the number $0.5W$ or $2W$. You now must decide whether to accept W dollars as your winnings or to switch. In this case, you would do better to switch.

Chapter 3. Reasoning across Domains

3.1 Statistics, Probability, and Inference

3.1.1. See the table on the next page.

3.1.2. In the list below, Y signifies that $m_x^2 = m_y$, and N means that the equality is not generally true.

    a. N   **b.** N   **c.** Y if n is odd. N if n is even.
    d. Y   **e.** Y   **f.** Y
    g. N   **h.** N   **i.** N
    j. N   **k.** Y   **l.** Y
   m. N

3.1.4.   **a.** By Tchebycheff's inequality, the probability is not greater than 0.69.

    b. 0.23

    c. 0.016

    d. Approximately 0.0000006.

Measures Characterizing Three Distributions of Scores

Measure	Distribution original scores x	adjusted scores scheme **a** $x + 5$	scheme **b** $1.15x$
Linear correlation coefficient with scores in test B (i.e., r)	0.55	0.55	0.55
Rank-order correlation coefficient with scores in test C (i.e., r_s)	0.61	0.61	0.61
Arithmetic mean (M)	58	63	66.7
Standard deviation (σ)	12	12	13.8
Variance (σ^2)	144	144	190.44
Median (Me)	59	64	67.85
Standard score of student with the lowest achievement ($z_{\min}$)	-2.8	-2.8	-2.8
Range (R)	64	64	73.6
Mode (Mo)	60	65	69
Percentile of student i (i.e., F_i)	78%	78%	78%

3.1.5.   **a.** 0.080

b. 0.770

c. 0.954

Note how the probability of achieving a given accuracy (± 2) in point estimation of the population's mean increases with the increase in sample size. Since, the greater the sample size the smaller the variance of the sampling distribution around μ_x, the mean of a larger random sample is more likely to be in a given neighborhood of μ_x.

3.1.6.   **a.**
$$
\begin{array}{lllll}
(\,1) & = (\,4) & = (\,7) & > (\,10) & > (\,13) & < \\
(\,2) & > (\,5) & < (\,8) & > (\,11) & < (\,14) & < \\
(\,3) & < (\,6) & = (\,9) & > (\,12) & = (\,15) &
\end{array}
$$

    b. (1) $<$

       (2) $=$

       (3) $<$

       (4) $>$

3.1.7.

	Population			
	Expectation 80; variance 16; shape of distribution unknown		Normal distribution of x; expectation: $\mu_x = 120$; variance (σ_x^2) unknown	
Statistic	M_{250}	$\dfrac{M_{400} - 80}{\sqrt{16/400}}$	$\dfrac{M_{18} - 120}{\sqrt{\dfrac{\sum_{i=1}^{18}(x_i - M_{18})^2}{17 \times 18}}}$	Number of observations exceeding 120 in a sample of size 12: $N(x \mid x > 120)$
Shape of sampling distribution	Approximately normal	Approximately normal	t distribution with 17 degrees of freedom	Binomial with $n = 12$; $p = 0.5$
Expectation of sampling distribution	80	0	0	6
Variance of sampling distribution	$\dfrac{16}{250} = 0.064$	1	$\dfrac{17}{15} = 1.13$	$6 \times 0.5 \times 0.5 = 1.5$

3.1.8.

	Questions about the distribution		
Random Variable	Shape	Expectation	Variance
$\overline{X}_{27} + \overline{Y}_{25}$	normal	180	7
$2 - 3\overline{X}_{36}$	normal	-238	20.25
$\dfrac{(\overline{Y}_{10} - \overline{X}_9) - 20}{\sqrt{\frac{100}{10} + \frac{81}{9}}}$	normal	0	1
$\left(\dfrac{X - 80}{9}\right)^2 + \left(\dfrac{Y - 100}{10}\right)^2$	chi-square distribution with two degrees of of freedom	2	4
The proportion of Y values exceeding 108.4 in a random sample of size 500: $\dfrac{N(Y \mid Y > 108.4)}{500}$	approximately normal	0.20	$\dfrac{0.20 \times 0.80}{500}$ $= 0.00032$

3.1.9.   **a.** $1 - 2 \times \left(\frac{1}{2}\right)^8 = 1 - \frac{1}{128} = 0.992.$

b. $\dbinom{8}{4} \left(\frac{1}{2}\right)^8 = \frac{35}{128} = 0.273.$

3.1.10.   **a.** $\alpha = P(R \mid H_0) = \dfrac{1}{\dbinom{7}{2}} = 0.048.$

b.

$$P(H_0 \mid R) = \frac{P(R \mid H_0)P(H_0)}{P(R \mid H_0)P(H_0) + P(R \mid H_1)P(H_1)}$$

$$= \frac{\dfrac{1}{\dbinom{7}{2}} \times 0.9}{\dfrac{1}{\dbinom{7}{2}} \times 0.9 + \dfrac{\dbinom{5}{2}}{\dbinom{7}{2}} \times 0.1} = 0.474.$$

Note that although R is relatively rare when H_0 is true, and the occurrence of R would entail *rejection* of H_0 for level of significance 0.05, in more than 47% of the cases in which R occurs, H_0 is still true. $P(H_0 \mid R)$ is almost 10 times greater than $P(R \mid H_0)$. These results demonstrate that when you get a significant result and ask yourself what is the probability that rejection of H_0 was an error, the answer is *not* equal to the level of significance of that test (see also Problem 6.1.16 in Part II); the Bayesian posterior probability of H_0 may still be quite high.

3.1.11.   **a.** 0.41

Method of solution. Let G denote the event that a **G**reen cab was the **G**uilty one. Obviously, $\overline{G}$ should be denoted B. Let g designate the event that the witness testifies that the hit-and-run cab was **g**reen.

In the absence of any information indicating otherwise, we assume that $P(G) = 0.15$ (this seems the fairest assumption for the *prior* probability of the target uncertain event). It is given that $P(g \mid G) = 0.80$ and $P(g \mid B) = 0.20$. Now, applying Bayes' theorem, we obtain the *posterior* probability of G:

$$P(G \mid g) = \frac{P(g \mid G)P(G)}{P(g \mid G)P(G) + P(g \mid B)P(B)}$$

$$= \frac{0.80 \times 0.15}{0.80 \times 0.15 + 0.20 \times 0.85} = 0.41 \ .$$

b. For the sake of the analysis in this section, let us denote the testimony of the first witness by g_1 and that of the second witness by g_2. (We assume independence between the statements given by the two witnesses, conditioned on either G or B.) The following are the calculations and the conclusions of the judges:

(a) Judge C:

$P(g_1 \cap g_2 \mid B) = 0.20 \times 0.20 = 0.04 < 0.05$.
The green company is found guilty.

(b) Judge B: $P(G \mid g_1 \cap g_2) =$

$$\frac{0.80 \times 0.80 \times 0.15}{0.80 \times 0.80 \times 0.15 + 0.20 \times 0.20 \times 0.85}$$

$$= 0.74 < 0.95 \ .$$

The green company is found not guilty.

Discussion. In some way, each of the judges is right. Let's look at a modified version of the original story problem, framed by Ginossar and Trope (1987) as a game of chance:

> Imagine an urn containing colored marbles. Suppose 15% of the marbles are green and 85% are blue. A person draws a marble randomly from the urn, looks at it, and states that it is green. An eye examination of this person showed that he suffers from color blindness and that he correctly identifies green and blue 80% of the time and fails 20% of the time (p. 471).

On the basis of this information, you are asked to judge the probability that the marble was green. It appears that the above calculation (in **a**) is justified for this version, so that a proportion of 0.41 out of all cases in which the witness testifies "g" are actually green. The same is true for the Bayesian calculation in section **b** if we extend the modified version to encompass the two witnesses. Assuming initial random sampling and regarding probabilities as relative frequencies (out of many repetitions of such accidents and testimonies), the answer of Judge B (0.74) is correct.

There is more to say in favor of view B. It considers all the available information: the base-rate frequencies and both likelihoods (under G and under B) of obtaining the testimonies, whereas argument C takes just the likelihood of the testimonies under B into consideration. In addition, Judge B answers the question of guilt of the green company by directly computing the greens' probability of guilt conditioned on all the available information. Judge C, in contrast, answers a question nobody has asked, namely, what is the probability of obtaining the double testimony, given the guilt of the blues?

On the other hand, the reasoning of Judge C seems right in that it deals only with the individual case at hand. If the legal system strives to do justice to the individual whom it tries, it should not introduce population frequencies into the judgment. Base-rates appear irrelevant to finding out the truth about the *unique* event judged.

Incorporating distributional information when weighing the evidence may paradoxically result in judging by frequencies. Suppose the first witness had testified "b." In that case, the Bayesian computation (relying on the same base rates) would result in $P(B \mid b) = 0.96$, exceeding Judge B's criterion for guilt. Were both witnesses to say a blue cab was involved in the accident, the Bayesian calculation would yield $P(B \mid b_1 \cap b_2) = 0.99$, certainly deeming the blues guilty. Thus, based on the same witnesses, Judge B would find the greens not guilty if the witnesses testify "g," and the blues guilty if they testify "b," although the witnesses are equally credible in both cases.

Carrying the situation to absurdity, a suspect cab company may come out guilty because its color prevails in town, and get exonerated, on the power of the same evidence, just because it is of a minority color. Ruling consistently by such a Bayesian rationale, a judge may justly be accused of prejudice because of ostensibly passing judgment by (skin) color.

In conclusion, the question of deciding between a classic and a Bayesian analysis of the cab problem will have to remain open.

3.1.12. $1 - (1 - 0.05)^{10} = 0.40$.

3.1.13.   **a.** (1) 0.0025
   (2) 0.0975
   (3) 0.095
   (4) 0.9025
   (5) 0.05

   b. (1) 0.64
   (2) 0.96
   (3) 0.32
   (4) 0.04
   (5) 0.80

3.1.14.   **a.**

Probabilities of All Possible Genotypes, with Inbreeding Coefficient I: Two Alleles

Value of sperm: Y	Value of egg: X		
	0	1	Total
1	$(1-I)pq$	$Ip + (1-I)p^2$	p
0	$Iq + (1-I)q^2$	$(1-I)pq$	q
Total	q	p	1

Note that when $I = 1$, the population is completely homozygous: it comprises only the genotypes 1,1 and 0,0 with respective probabilities p and q. When $I = 0$, the probabilities reduce to those obtained in case of independence: every joint probability equals the product of the corresponding marginal probabilities. When considering the probabilities of the two heterozygous genotypes, we see that $1 - I$ is the multiplicative factor by which heterozygosity is changed relative to the case of independence. Thus, the inbreeding coefficient, I, measures the *fraction by which heterozygosity is reduced* (Crow & Kimura, 1970, p. 66).

b. Based on the joint probability distribution in **a**, we easily obtain

$$E(X) = E(Y) = p$$

$$\sigma^2(X) = \sigma^2(Y) = pq = p(1-p)$$

$$E(XY) = Ip + (1-I)p^2.$$

Using the formula

$$r = \frac{E(XY) - E(X)E(Y)}{\sigma(X)\sigma(Y)},$$

we get

$$r = \frac{Ip + (1-I)p^2 - p^2}{p(1-p)} = \frac{Ip - Ip^2}{p(1-p)} = I.$$

This completes the proof.

Note that the inbreeding coefficient, I, defined as a probability, is a number between 0 and 1, whereas the correlation coefficient is bounded by -1 and $+1$. This means that only r values between 0 and 1 can equal the probability of identity by descent.

The above equality of r and I, suggests a new interpretation of a *nonnegative correlation coefficient* as the *probability of "identity by descent."* The interpretation as the "fraction by which heterozygosity is reduced" (see answer to **a**) can, however, be extended to the whole range of r values. A negative correlation coefficient (as well as a negative inbreeding coefficient) would signify an increase, instead of a decrease, in heterozygosity ('outbreeding' instead of inbreeding).

c. We first compute the expectations and variances of the marginal distributions:

$$E(X) = E(Y) = p$$

$$\sigma^2(X) = \sigma^2(Y) = pq = p(1-p) .$$

Second, we compute the expectation of the variable XY. The result $E(XY) = p_{11}$ is immediate.

Inserting the results of these computations into the formula for r, we obtain,

$$r = \frac{p_{11} - p^2}{p(1-p)} .$$

We can now solve for p_{11} to obtain

$$p_{11} = rp + (1-r)p^2 .$$

Note that p_{11} turned out the same as the probability of genotype 1,1 in **a**, with I replaced by r. Since a 2×2 distribution with fixed marginals has only one degree of freedom, each of the four probabilities must equal the probability of the respective genotype in **a**, with r taking the place of I. Thus, the correlation coefficient of *any* 2×2 distribution with equal marginals plays the role of the inbreeding coefficient in the case of two alleles.

This result means that when $r \geq 0$, every 2×2 distribution with equal marginals can be formally construed to be composed of (proportion r) cases of 'identical descent,' that is, cases that are *inherently identical*; and (proportion $1-r$) cases which are independently distributed, that is, randomly paired. The degree to which this interpretation is meaningful depends on the particular context and the definition of the variables.

More generally, for $-1 \leq r \leq +1$, the correlation coefficient of any 2×2 distribution with equal marginal distributions can be interpreted as the *fraction by which inequality is decreased*, where negative r's refer to increases in inequality between the variables. It is important to note that this decrease (or increase) in inequality is computed as a fraction of the rate of inequality in case of independence (i.e., when $r = 0$). This adds another interpretation to a long list of interpretations of the correlation coefficient (see, e.g., Ozer, 1985; Rodgers & Nicewander, 1988).

d. We have
$$p_{ij} = (1 - I)p_i p_j + \delta_{ij} I p_i,$$
where $\delta_{ij} = \left\{ \begin{array}{ll} 1 & \text{if} \quad i = j \\ 0 & \text{if} \quad i \neq j \end{array} \right\}$ for $i, j = 1, 2, \ldots, n$.

As mentioned, a joint distribution satisfying this formula is labeled a *genetic distribution*.

e. Using the notations introduced in **d**, we first note that because of the equal marginal distributions,
$$E(X) = E(Y) = E(A),$$
and
$$\sigma^2(X) = \sigma^2(Y) = \sigma^2(A).$$

Second, we compute $E(XY)$, relying on the formula for the genetic distribution:

$$E(XY) = (1 - I) \sum_j \sum_i p_i p_j a_i a_j + I \sum_j \sum_i \delta_{ij} p_i a_i a_j$$

$$= (1 - I) \sum_j p_j a_j \sum_i p_i a_i + I \sum_i p_i a_i^2$$

$$= (1 - I) E(A) E(A) + I E(A^2)$$

$$= [E(A)]^2 + I[E(A^2) - (E(A))^2]$$

$$= [E(A)]^2 + I \sigma^2(A) .$$

Substituting these values in the formula for r, we obtain,

$$r = \frac{[E(A)]^2 + I \sigma^2(A) - [E(A)]^2}{\sigma^2(A)} = I .$$

This completes the proof.

Given a genetic distribution, we can extend the interpretation of a nonnegative r as the probability of originating from common descent to the case of n alleles. Likewise, we can interpret any r (in the range $-1 \le r \le +1$) as the fraction by which heterozygosity is reduced. This is true regardless of the value of each allele and the number of alleles (Crow & Kimura, 1970, pp. 66–68).

In the opposite direction, however, the conclusion of the 2×2 case cannot, as a rule, be extended to the general ($n \times n$) case. An arbitrary bivariate distribution with equal marginal distributions, and a given r, *cannot* always be construed as the joint distribution of the values of paired alleles with inbreeding coefficient r.

Thus, a joint $n \times n$ probability distribution, with equal marginal distributions, is not necessarily a genetic distribution in which r plays the role of the inbreeding coefficient, except for the case where $n = 2$. This means that when $n > 2$, a bivariate distribution is no longer uniquely determined by the conjunction of the marginal probabilities and r. It is easy to see why this is so if we consider the fact that a genetic distribution is symmetric, whereas an arbitrary two-dimensional distribution, even though its marginals may be equal, need not be symmetric.[16] Hence, the validity of the interpretation of a nonnegative r as the probability of "identity by descent" is limited to symmetric 2×2 distributions and to $n \times n$ genetic distributions.

---

[16]The reader is advised to make up one counterexample. It could be a 3×3 probability distribution, with equal marginal distributions, that is *not* symmetric (and consequently, is not a genetic distribution).

Answers for Part II: Multiple-Choice Problems

Chapter 4. Descriptive Statistics II

4.1 Scales of Measurement

4.1.1. c.
4.1.2. a.
4.1.3. d.
4.1.4. g.
4.1.5. a.
4.1.6. b.
4.1.7. b.
4.1.8. c.

4.2 Measures Characterizing Distributions II

4.2.1. e.
4.2.2. d.
4.2.3. c.
4.2.4. c.
4.2.5. b.
4.2.6. a.
4.2.7. g.
4.2.8. f.
4.2.9. e.

4.2.10. d.
4.2.11. f.
4.2.12. e.
4.2.13. d.
4.2.14. c.
4.2.15. a.
4.2.16. e.
4.2.17. g.
4.2.18. d.

4.3 Transformed Scores, Association, and Linear Regression

4.3.1.	a.	**4.3. 9.**		c.
4.3.2.	b.	**4.3.10.**		f.
4.3.3.	a.	**4.3.11.** (1)		d.
4.3.4.	f.	**4.3.11.** (2)		c.
4.3.5.	d.	**4.3.12.** (1)		d.
4.3.6.	a.	**4.3.12.** (2)		d.
4.3.7.	b.	**4.3.13.**		a.
4.3.8.	b.	**4.3.14.**		d.

4.4 In Retrospect

4.4.1 b.

Chapter 5. Probability II

5.1.1.	b.	**5.1.11.**	b.
5.1.2.	c.	**5.1.12.**	b.
5.1.3.	e.	**5.1.13.**	e.
5.1.4.	d.	**5.1.14.**	e.
5.1.5.	e.	**5.1.15.**	c.
5.1.6.	a.	**5.1.16.**	f.
5.1.7.	c.	**5.1.17.**	c.
5.1.8.	f.	**5.1.18.**	e.
5.1.9.	d.	**5.1.19.**	d.
5.1.10.	d.		

Chapter 6. Normal Distribution, Sampling Distributions, and Inference

6.1.1.	g.	**6.1. 9.**	e.
6.1.2.	f.	**6.1.10.**	b.
6.1.3.	d.	**6.1.11.**	e.
6.1.4. (1)	a.	**6.1.12.**	b.
6.1.4. (2)	b.	**6.1.13.**	d.
6.1.5.	d.	**6.1.14.**	a.
6.1.6.	e.	**6.1.15.**	d.
6.1.7.	e.	**6.1.16.**	f.
6.1.8.	d.	**6.1.17.**	a.

References *

Arbel, T. (1985). Minimizing the sum of absolute deviations. *Teaching Statistics*, **7**, 88–89. [139]

Armstrong, R. (1982). An area model for solving probability problems. In R. D. Armstrong & P. Pedersen (Eds.), *Probability and statistics: A collection of papers on the teaching of probability and statistics in CSMP's elementary school curriculum* (pp. 77–88). St. Louis, MO: CEMREL. [42, 43]

Ayton, P., & McClelland, A. (1987). The despicable Doctor Fischer's (Bayesian) bomb party. *Teaching Mathematics and its Applications*, **6**, 179–183. [56, 180]

Ballinger, D. G., & Benzer, S. (1989). Targeted gene mutations in Drosophila. *Proceedings of the National Academy of Sciences, Washington*, **86**, 9402–9406. [63]

Bar-Hillel, M., & Falk, R. (1982). Some teasers concerning conditional probabilities. *Cognition*, **11**, 109–122. [56, 186]

Bickel, P.J., Hammel, E.A., & O'Connell, J.W. (1975). Sex bias in graduate admissions: Data from Berkeley. *Science*, **187**, 398–404. [41]

Book, S. A., & Sher, L. (1979). How close are the mean and the median? *Two-Year College Mathematics Journal*, **10**, 202–204. [15]

Borovcnik, M. G. (1988). Revising probabilities according to new information: A fundamental stochastic intuition. In R. Davidson & J. Swift (Eds.), *Proceedings of the second international conference on teaching statistics*. Victoria B.C.: University of Victoria. [168]

---

* The numbers in brackets indicate the page numbers where the item is cited in the text.

Bullen, P. S. (1990). Averages still on the move. *Mathematics Magazine*, **63**, 250–255. [138]

Burrows, B. L., & Talbot, R. F. (1986). Which mean do you mean? *International Journal of Mathematical Education in Science and Technology*, **17**, 275–284. [14, 138]

Bush, R. R., & Mosteller, F. (1955). *Stochastic models for learning*. New York: Wiley. [x]

Butchart, J. H., & Moser, L. (1952). No calculus, please. *Scripta Mathematica*, **18**, 221–236. [139]

Campbell, C., & Joiner, B. L. (1973). How to get the answer without being sure you've asked the question. *The American Statistician*, **27**, 229–231. [40]

Carver, R. P. (1978). The case against statistical significance testing. *Harvard Educational Review*, **48**, 378–399. [79]

Chaudhuri, A., & Mukerjee, B. (1988). *Randomized response: Theory and technique*. New York: Marcel Dekker. [40]

Chow, S. L. (1988). Significance test or effect size? *Psychological Bulletin*, **103**, 105–110. [xiii, 128, 129]

Chu, D., & Chu, J. (1992). A "simple" probability problem. *Mathematics Teacher*, **85**, 191–195. [56]

Chung, K. L. (1942). On mutually favorable events. *Annals of Mathematical Statistics*, **13**, 338–349. [45]

Crow, J. F., & Kimura, M. (1970). *An introduction to population genetics theory*. New York: Harper & Row. [81, 82, 211, 214]

Dessart, D. J. (1971). To tip a waiter — A problem in unordered selections with repetitions. *Mathematics Teacher*, **64**, 307–310. [154]

Devore, J. L. (1979). Estimating a population proportion using randomized responses. *Mathematics Magazine*, **52**, 38–40. [40]

Diaconis, P., & Efron, B. (1983). Computer-intensive methods in statistics. *Scientific American*, **248**, 96–108. [xi]

Diaconis, P., & Mosteller, F. (1989). Methods for studying coincidences. *Journal of the American Statistical Association*, **84**, 853–861. [31]

Diaconis, P., & Zabell, S. (1986). Some alternatives to Bayes's rule. In B. Grofman & G. Owen (Eds.), *Information and group decision making: Proceedings of the second University of California, Irvine, conference on political economy* (pp. 25–38). Greenwich, CT: JAI Press. [187]

Eddington, A. S. (1935). The problem of *A, B, C,* and *D. The Mathematical Gazette,* **19**, 256–257. [55]

Efron, B., & Tibshirani, R. (1991). Statistical data analysis in the computer age. *Science,* **253**, 390–395. [xi]

Eisenbach, R. (1988). *Statistics for "nonstatisticians"* (in Hebrew). Jerusalem: Akademon. [90]

Falk, R. (1979a). Revision of probabilities and the time axis. *Proceedings of the third international conference for the psychology of mathematics education* (pp. 64–66). Warwick, England. [167]

Falk, R. (1979b). 3:40 mile. *Mathematics Teacher,* **72**, 489. [144]

Falk, R. (1980). Minimise your losses. *Teaching Statistics,* **2**, 80–83. [6, 7]

Falk, R. (1981). Another look at the mean, median, and standard deviation. *Two-Year College Mathematics Journal,* **12**, 207–208. [15]

Falk, R. (1982). Do men have more sisters than women? *Teaching Statistics,* **4**, 60–62. [45]

Falk, R. (1983). Probabilistic reasoning as an extension of commonsense thinking. In M. Zweng, T. Green, J. Kilpatrick, H. Pollak, & M. Suydam (Eds.), *Proceedings of the fourth international congress on mathematical education* (pp. 190–195). Boston: Birkhäuser. [52]

Falk, R. (1984). Multiplicative analogues of some statistics. *The American Mathematical Monthly,* **91**, 198–202. [138]

Falk, R. (1986a). Misconceptions of statistical significance. *Journal of Structural Learning,* **9**, 83–96. [77]

Falk, R. (1986b). Of probabilistic knights and knaves. *The College Mathematics Journal,* **17**, 156–164. [55]

Falk, R. (1986c). Some distributions and expectations — simplified. *International Journal of Mathematical Education in Science and Technology,* **17**, 487–495. [65, 66, 193]

Falk, R. (1989). Inference under uncertainty via conditional probabilities. In R. Morris (Ed.), *Studies in mathematics education, Vol. 7: The teaching of statistics* (pp. 175–184). Paris: Unesco. [167, 178]

Falk, R. (1992). A closer look at the probabilities of the notorious three prisoners. *Cognition*, **43**, 197–223. [59, 187]

Falk, R., & Bar-Hillel, M. (1980). Magic possibilities of the weighted average. *Mathematics Magazine*, **53**, 106–107. [12, 41, 64]

Falk, R., & Bar-Hillel, M. (1983). Probabilistic dependence between events. *Two-Year College Mathematics Journal*, **14**, 240–247. [45]

Falk, R., & Konold, C. (1992). The psychology of learning probability. In F. S. Gordon & S. P. Gordon (Eds.), *Statistics for the twenty-first century* (pp. 151–164). USA: Mathematical Association of America. [45, 56, 67]

Falk, R., Lipson, A., & Konold, C. (in press). The ups and downs of the hope function in a fruitless search. In G. Wright & P. Ayton (Eds.), *Subjective probability*. Chichester, Wiley. [182]

Falk, R., & Well, A. (in preparation). The many faces of the correlation coefficient. Unpublished manuscript. The Hebrew University of Jerusalem and University of Massachusetts, Amherst. [81]

Feller, W. (1957). *An introduction to probability theory and its applications*. Vol. 1 (2nd ed.). New York: Wiley. [62, 154, 156, 177, 193, 194, 196, 198]

Fisher, R. A. (1960). *The design of experiments* (7th ed.). Edinburgh: Oliver & Boyd. [28]

Fox, J. A., & Tracy, P. E. (1986). *Randomized response: A method for sensitive surveys*. London: Sage. [40]

Freedman, D., Pisani, R., & Purves, R. (1978). *Statistics*. New York: Norton. [10, 11, 30, 39, 149]

Freudenthal, H. (1970). The aims of teaching probability. In L. Råde (Ed.), *The teaching of probability and statistics* (pp. 151-167). Stockholm: Almqvist & Wiksell. [39, 65, 196]

Galton, F. (1869). *Hereditary genius: An inquiry into its laws and consequences*. London: Macmillan. [164]

Galton, F. (1894). A plausible paradox in chances. *Nature*, **49**, 365–366. [27]

Galton, F. (1904). Average number of kinsfolk in each degree. *Nature*, **70**, 529 & 626. [164]

Gardner, M. (1957). *Fads and fallacies in the name of science.* New York: Dover. [36]

Gardner, M. (1961). *The second Scientific American book of mathematical puzzles and diversions.* New York: Simon & Schuster. [187]

Gardner, M. (1982). *Aha! Gotcha: Paradoxes to puzzle and delight.* New York: Freeman. [15, 62, 138, 202]

Gardner, M. (1992). Probability paradoxes. *Skeptical Inquirer*, **16**, 129–132. [59, 187, 188]

Gigerenzer, G., Swijtink, Z., Porter, T., Daston, L., Beatty, J., & Krüger, L. (1989). *The empire of chance: How probability changed science and everyday life.* Cambridge: Cambridge University Press. [186]

Gillman, L. (June-July 1991). The car-and-goats fiasco. *Focus: The Newsletter of the Mathematical Association of America*, **11**(3), 8. [59, 187]

Gillman, L. (1992). The car and the goats. *The American Mathematical Monthly*, **99**, 3–7. [59, 187]

Ginossar, Z., & Trope, Y. (1987). Problem solving in judgment under uncertainty. *Journal of Personality and Social Psychology*, **52**, 464–474. [208]

Glickman, L. V. (1982). Families, children, and probabilities. *Teaching Statistics*, **4**, 66–69. [56, 178]

Glickman, L. (1989). Why teach the history of probability? *Teaching Statistics*, **11**, 6–7. [39]

Gnanadesikan, M., Scheaffer, R., & Swift, J. (1987). *The art and techniques of simulation.* Palo Alto, CA: Dale Seymour Publications. [xi]

Golomb, S. W. (1968). New proof of a classic combinatorial theorem. *The American Mathematical Monthly*, **75**, 530–531. [154]

Greene, G. (1980). *Doctor Fischer of Geneva or the bomb party.* London: Penguin. [56, 57]

Gudder, S. (1981). Do good hands attract? *Mathematics Magazine*, **54**, 13–16. [45]

Gupta, H. (1955). Distributing m similar objects among n persons (Problem E 1163. Including solutions by J. V. Pennington and J. Braun). *The American Mathematical Monthly*, **62**, 730–731. [154]

Guttman, L. (1977). What is not what in statistics. *The Statistician*, **26**, 81–107. [79]

Hacking, I. (1975). *The emergence of probability*. Cambridge: Cambridge University Press. [186]

Hadar, N., & Hadass, R. (1981a). Between associativity and commutativity. *International Journal of Mathematical Education in Science and Technology*, **12**, 535–539. [93]

Hadar, N., & Hadass, R. (1981b). The road to solving a combinatorial problem is strewn with pitfalls. *Educational Studies in Mathematics*, **12**, 435–443. [28]

Hamdan, M. A. (1978). A systematic approach to teaching counting formulae. *International Statistical Review*, **46**, 219–220. [152]

Hays, W. L., & Winkler, R. L. (1971). *Statistics: Probability, inference and decision*. New York: Holt, Rinehart & Winston. [193, 194, 198]

Hering, F. (1987). A new version of playing battleships. *International Journal of Mathematical Education in Science and Technology*, **18**, 417–432. [49]

Hocking, R. L., & Schwertman, N. C. (1986). An extension of the birthday problem to exactly k matches. *The College Mathematics Journal*, **17**, 315–321. [31]

Hogg, R. V., & Tanis, E. A. (1977). *Probability and statistical inference*. New York: Macmillan. [193, 198, 200]

Hoehn, L., & Niven, I. (1985). Averages on the move. *Mathematics Magazine*, **58**, 151–156. [138]

Holm, S. (1970). A collection of problems in probability and statistics. In L. Råde (Ed.), *The teaching of probability and statistics* (pp. 347–357). Stockholm: Almqvist & Wiksell. [45]

Honsberger, R. (1978). *Mathematical morsels*. USA: Mathematical Association of America. [140]

Hooke, R. (1983). *How to tell the liars from the statisticians.* New York: Marcel Dekker. [183]

Joag-Dev, K. (1989). MAD property of a median: A simple proof. *The American Statistician,* **43**, 26–27. [139]

Kahneman, D., & Tversky, A. (1982). Variants of uncertainty. *Cognition,* **11**, 143–157. [167]

Kasner, E., & Newman, J. (1949). *Mathematics and the imagination.* London: Bell & Sons. [36]

Kokan, A. R. (1975). The minimum property of the mean deviation. *The Mathematical Gazette,* **59** (No.408), 111. [139]

Konold, C. (1989). Informal conceptions of probability. *Cognition & Instruction,* **6**, 59–98. [111]

Konold, C. (in press). An example of teaching probability through modeling real problems. *Mathematics Teacher.* [47]

Kreith, K. (1976). Mathematics, social decisions and the law. *International Journal of Mathematical Education in Science and Technology,* **7**, 315–330. [40]

Kreith, K. (July/August 1992). Summertime, and the choosin' ain't easy: An ice cream counting problem. *Quantum,* 28–30. [32]

Kreith, K., & Kysh, J. (1988). The fourth way to sample *k* objects from a collection of *n*. *Mathematics Teacher,* **81**, 146–149. [30, 152, 154]

Lichtenstein, S., Fischhoff, B., & Phillips, L. D. (1982). Calibration of probabilities: The state of the art to 1980. In D. Kahneman, P. Slovic, & A. Tversky (Eds.), *Judgment under uncertainty: Heuristics and biases* (pp. 306–334). Cambridge: Cambridge University Press. [111]

Lightner, J. E. (1991). A brief look at the history of probability and statistics. *Mathematics Teacher,* **84**, 623–630. [27]

Loyer, M. W. (1983). Bad probability, good statistics, and group testing for binomial estimation. *The American Statistician,* **37**, 57–59. [178]

Lubecke, A. M. (1991). Which mean do you mean? *Mathematics Teacher,* **84**, 24–28. [14]

Madsen, R. W. (1981). Making students aware of bias. *Teaching Statistics,* **3**, 2–5. [13]

Maor, E. (1977). A mathematician's repertoire of means. *Mathematics Teacher*, **70**, 20–25. [138]

Meehl, P. E. (1956). Wanted — a good cookbook. *The American Psychologist*, **11**, 263–272. [55,176]

Meshalkin, L. D. (1973). *Collection of problems in probability theory* (L. F. Boron & B. A. Haworth, Trans.). Leyden: Noordhoff International. (Original work published 1963.) [57, 156]

Morgan, J. P., Chaganty, N. R., Dahiya, R. C., & Doviak, M. J. (1991). Let's make a deal: The player's dilemma. *The American Statistician*, **45**, 284–287. [59]

Mosteller, F. (1965). *Fifty challenging problems in probability with solutions*. Reading, MA: Addison-Wesley. [31,62, 187]

Mosteller, F. (1980). Classroom and platform performance. *The American Statistician*, **34**(1), 11–17. [45]

Noether, G. E. (1980). The role of nonparametrics in introductory statistics courses. *The American Statistician*, **34**(1), 22–23. [76]

Oakley, C. O., & Baker, J. C. (1977). Least squares and the 3:40-minute mile. *Mathematics Teacher*, **70**, 322–324. [144]

O'Beirne, T. H. (1961). The truth about the four liars. *New Scientist* (No. 234), 330–331. [55]

Ozer, D. J. (1985). Correlation and the coefficient of determination. *Psychological Bulletin*, **97**, 307–315. [213]

Page, W., & Murty, V. N. (1982). Nearness relations among measures of central tendency and dispersion: Part 1. *Two-Year College Mathematics Journal*, **13**, 315–327. [15, 141]

Page, W., & Murty, V. N. (1983). Nearness relations among measures of central tendency and dispersion: Part 2. *Two-Year College Mathematics Journal*, **14**, 8–17. [15, 141]

Pauker, S. P., & Pauker, S. G. (1979). The amniocentesis decision: An explicit guide for parents. In C.J. Epstein, C.J.R. Curry, S. Packman, S. Sherman, & B.D. Hall (Eds.), *Birth defects: Original article series. Vol 15: Risk, communication, and decision making in genetic counseling* (pp. 289–324). New York: The National Foundation. [54]

Paulos, J. A. (1988). *Innumeracy: Mathematical illiteracy and its consequences.* New York: Hill & Wang. [62, 63]

Paulson, R. A. (1992). Using lottery games to illustrate statistical concepts and abuses. *The American Statistician*, **46**, 202–204. [47]

Polya, G. (1957). *How to solve it: A new aspect of mathematical method* (2nd ed.). Princeton, NJ: Princeton University Press. [x]

Rodgers, J. L., & Nicewander, W. A. (1988). Thirteen ways to look at the correlation coefficient. *The American Statistician*, **42**, 59–66. [213]

Roughgarden, J. (1979). *Theory of population genetics and evolutionary ecology: An introduction.* New York: Macmillan. [81]

Saunders, S. C. (April 1990). The shell game, prisoner paradox and other paradoxes of conditional probability. *Mathematics Notes from Washington State University*, **33**, No. 2 (Whole Number 129). [59]

Schwertman, N. C., Gilks, A. J., & Cameron, J. (1990). A simple noncalculus proof that the median minimizes the sum of absolute deviations. *The American Statistician*, **44**, 38–39. [139]

Selvin, S. (1975a). A problem in probability. *The American Statistician*, **29**, 67. [59]

Selvin, S. (1975b). On the Monty Hall problem. *The American Statistician*, **29**, 134. [59]

Shaughnessy, J. M., & Dick, T. (1991). Monty's dilemma: Should you stick or switch? *Mathematics Teacher*, **84**, 252–256. [59]

Sher, L. (1979). The range of the standard deviation. *Two-Year College Mathematics Journal*, **10**, 33. [141]

Shimojo, S., & Ichikawa, S. (1989). Intuitive reasoning about probability: Theoretical and experimental analyses of the "problem of three prisoners." *Cognition*, **32**, 1–24. [187, 188]

Siegel, S. (1956). *Nonparametric statistics for the behavioral sciences.* New York: McGraw-Hill. [198, 200]

Siegel, S., & Castellan, N. J. Jr. (1988). *Nonparametric statistics for the behavioral sciences* (2nd ed.). New York: McGraw-Hill. [87]

Skemp, R. R. (1971). *The psychology of learning mathematics.* Harmondsworth, England: Penguin. [ix]

Stanley, R. P. (1986). *Enumerative Combinatorics.* Belmont, CA: Wadsworth. [152]

Székely, G. J. (1986). *Paradoxes in probability theory and mathematical statistics.* Dordrecht, Holland: Reidel. [39, 43, 187]

Troccolo, J. A. (1977). Randomness in physics and mathematics. *Mathematics Teacher*, **70**, 772–774. [33, 34, 152, 154, 156]

Tukey, J. W. (1977). *Exploratory data analysis.* Reading, MA: Addison-Wesley. [xi]

Tversky, A., & Kahneman, D. (1974). Judgment under uncertainty: Heuristics and biases. *Science*, **185**, 1124–1131. [123]

Tversky, A., & Kahneman, D. (1980). Causal schemas in judgments under uncertainty. In M. Fishbein (Ed.), *Progress in social psychology* (Vol. 1, pp. 49–72). Hillsdale, NJ: Erlbaum. [78, 169]

Tversky, A., & Kahneman D. (1983). Extensional versus intuitive reasoning: The conjunction fallacy in probability judgment. *Psychological Review*, **90**, 293–315. [22, 144]

Usiskin, Z. (1974). Some corresponding properties of real numbers and implications for teaching. *Educational Studies in Mathematics*, **5**, 279–290. [138]

Wagner, C. H. (1982). Simpson's paradox in real life. *The American Statistician*, **36**, 46–48. [42]

Walpole, R. E. (1974). *Introduction to statistics* (2nd ed.). New York: Macmillan. [193]

Walter, M. (1981). How to get a higher grade: Beware of averages. *Teaching Statistics*, **3**, 77–79. [93]

Williams, J. D. (1966). *The compleat strategyst: Being a primer on the theory of games of strategy.* New York: McGraw-Hill. [23]

Yule, G. U., & Kendall, M. G. (1953). *An introduction to the theory of statistics* (14th ed.). London: Charles Griffin. [139]

Zabell, S. L. (1988a). Loss and gain: The exchange paradox (discussion of a paper by B. M. Hill). In J. M. Bernardo, M. H. Degroot, D. V. Lindley, & A. F. M. Smith (Eds.), *Proceedings of the third Valencia international meeting* (pp. 233-236). Oxford: Clarendon Press. [67]

Zabell, S. L. (1988b). The probabilistic analysis of testimony. *Journal of Statistical Planning and Inference*, **20**, 327–354. [55, 187]

Zabell, S. L. (1988c). Symmetry and its discontents. In B. Skyrms & W. L. Harper (Eds.), *Causation, chance, and credence* (Vol. 1, pp. 155-190). Dordrecht, Holland: Kluwer Academic Publishers. [186]

Index

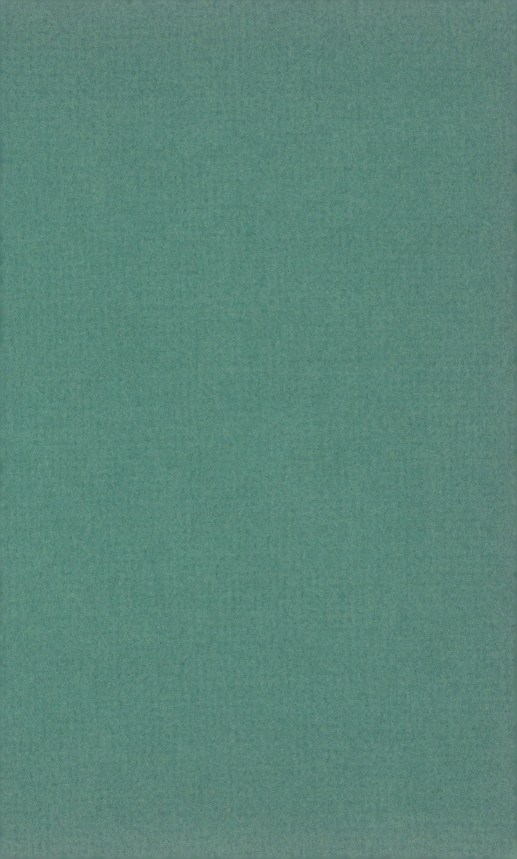

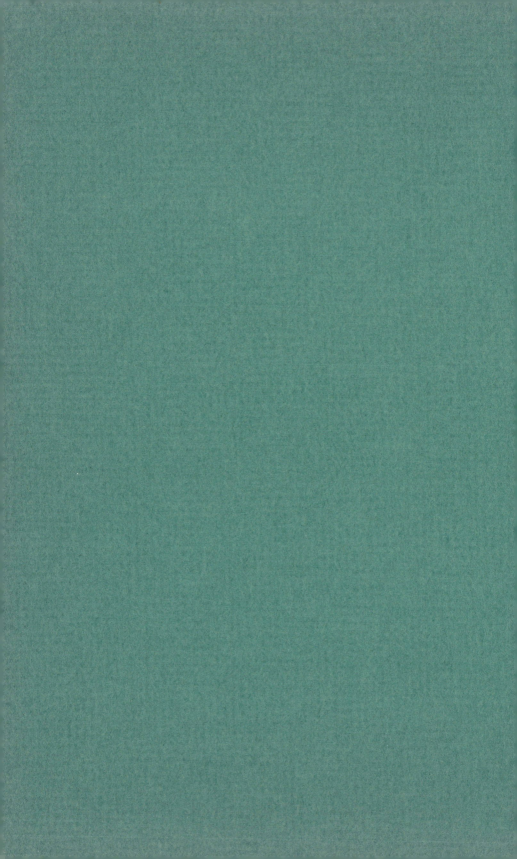